Advanced Topics in Science and Technology in China

Volume 5

Zhejiang University is one of the leading universities in China. In Advanced Topics in Science and Technology in China, Zhejiang University Press and Springer jointly publish monographs by Chinese scholars and professors, as well as invited authors and editors from abroad who are outstanding experts and scholars in their fields. This series will be of interest to researchers, lecturers, and graduate students alike.

Advanced Topics in Science and Technology in China aims to present the latest and most cutting-edge theories, techniques, and methodologies in various research areas in China. It covers all disciplines in the fields of natural science and technology, including but not limited to, computer science, materials science, the life sciences, engineering, environmental sciences, mathematics, and physics.

This book series is indexed by both the **Scopus** and **Compendex** databases.

If you are interested in publishing your book in the series, please contact Violetta Xu (Email: violetta_xuqian@zju.edu.cn) and Mengchu Huang (Email: mengchu.huang@cn.springer.com).

Dan Wu · Haotian Li · Qinfeng Zhu ·
Xingzhou Chen · Li-Ang Xu

Development and Application of 3D Diffusion Magnetic Resonance Imaging Sequences

ZHEJIANG UNIVERSITY PRESS
浙江大学出版社

Springer

Dan Wu
Zhejiang University
Hangzhou, Zhejiang, China

Haotian Li
Zhejiang University
Hangzhou, Zhejiang, China

Qinfeng Zhu
Zhejiang University
Hangzhou, Zhejiang, China

Xingzhou Chen
Zhejiang University
Hangzhou, Zhejiang, China

Li-Ang Xu
Zhejiang University
Hangzhou, Zhejiang, China

ISSN 1995-6819 ISSN 1995-6827 (electronic)
Advanced Topics in Science and Technology in China
ISBN 978-981-95-0807-5 ISBN 978-981-95-0808-2 (eBook)
https://doi.org/10.1007/978-981-95-0808-2

Jointly published with Zhejiang University Press
The print edition is not for sale in China (Mainland). Customers from China (Mainland) please order the print book from: Zhejiang University Press.

This Springer imprint is published by the registered company Springer Nature Singapore Pte Ltd.
The registered company address is: 152 Beach Road, #21-01/04 Gateway East, Singapore 189721, Singapore

Preface

Diffusion magnetic resonance imaging (dMRI) has been considered a primary non-invasive tool for studying the microstructure of in vivo tissue. Time-dependent (TD) dMRI can obtain tissue signals at multiple diffusion times, which is critical for understanding and depicting the microstructure of tissues. However, despite significant achievements in basic scientific research and clinical applications, clinical routine two-dimensional (2D) imaging still faces several major technical challenges, such as low signal-to-noise ratio (SNR) and resolution, long scanning time, phase errors in multi-shot acquisition, and brain tissue signals affected by cerebrospinal fluid (CSF). These issues are particularly severe in TDDMRI. In this book, we aim to introduce three-dimensional (3D) whole-brain TDDMRI technique with high SNR, high resolution, and high fidelity for multi-tap acquisition.

We mainly introduced a novel 3D DGprep-GRASE sequence that incorporated a series of new technologies, such as the multi-shot acquisition with the 3D navigator, the 3D GRAPPA, and the IR to address the technical challenges faced by clinical routine 2D imaging, such as low SNR and resolution, long scanning time, phase errors, and CSF influences. These technologies have significantly improved image SNR and resolution, reduced scan time, corrected phase errors between shots, and eliminated the influence of CSF. Therefore, the related research has achieved a new type of 3D whole-brain TDDMRI technique with high SNR, high resolution, and high fidelity for multi-shot acquisition, providing an important scientific basis for understanding and applying microstructural parameter analysis based on diffusion time-dependency, which further facilitates the clinical application for future clinical diagnosis and treatment.

While drafting this book, the students in our group translated and proofread some manuscripts. The preparation of this book drew on the knowledge of a number of classic textbooks in the field of magnetic resonance, which had been of great assistance. The editorial staff of Zhejiang University Press contributed to the publication and quality of this book, and we would like to express our heartfelt thanks to them.

Limited to the level of the editor and the preparation of the time in a hurry, the book was inevitable omissions or shortcomings, please readers did not hesitate to criticize and correct.

Hangzhou, China

Dan Wu
Haotian Li
Qinfeng Zhu
Xingzhou Chen
Li-Ang Xu

Contents

Chapter 1
Introduction

Abstract Magnetic Resonance Imaging (MRI) utilizes the spin properties of hydrogen nuclei in a strong magnetic field and the response of radiofrequency impulses to generate high-resolution images, and its non-invasive and multi-parametric imaging capabilities make it irreplaceable in the diagnosis of oncology, neurological and cardiovascular diseases. Diffusion Magnetic Resonance Imaging (dMRI) reveals the microstructure of tissues by measuring the diffusion motion of water molecules. For example, Diffusion Weighted Imaging (DWI) and Diffusion Tensor Imaging (DTI) quantify the direction of nerve fibers, etc., which provide critical information for stroke, tumors and neurodegenerative diseases. Microstructure modeling is an advanced method derived from conventional dMRI. By systematically varying diffusion-weighted parameters—such as gradient strength, direction, and diffusion time. It probes the behavior of water diffusion within biological microenvironments. It then applies mathematical and physical models to resolve key microstructural characteristics of tissue at the cellular level. This technique is broadly divided into two main strategies: q-space modeling, which manipulates diffusion gradient parameters, and t-space modeling, which alters diffusion time.

Keyword Magnetic resonance principle · Diffusion magnetic resonance imaging · Diffusion time dependence

1.1 Principle of Magnetic Resonance Imaging

MRI [1–3] is widely used in medical examinations due to its non-invasive nature. It provides detailed information about anatomical structures and functional connections through the principles of nuclear magnetic resonance and mathematical techniques such as Fourier transformation, offering a comprehensive view of the human body. The main benefit of MRI is its ability to produce high-resolution images with distinct contrast in soft tissues. Compared to other imaging techniques, such as X-rays [4] and computed tomography (CT) [5], MRI can more accurately depict tissue differences, providing a significant advantage in diagnosing various conditions, including tumors,

D. Wu et al., *Development and Application of 3D Diffusion Magnetic Resonance Imaging Sequences*, Advanced Topics in Science and Technology in China 5,
https://doi.org/10.1007/978-981-95-0808-2_1

neurological disorders, and cardiovascular diseases. Moreover, MRI offers diverse imaging modalities, allowing for adjustments in scanning sequences and parameters based on specific needs. This enables the acquisition of images that reflect different tissue characteristics, such as T1-weighted images [6], T2-weighted images [7], and diffusion-weighted images (DWI) [8]. These different modalities provide a wealth of information regarding the structure, function, and metabolic state of tissues. Therefore, MRI is extensively used in the medical field as a powerful tool for diagnosis and disease monitoring [9].

MRI possesses certain limitations that must be acknowledged. First, the cost of MRI machines is high, and operating them requires trained professionals. Second, the scanning process can be time-consuming, which may affect patient comfort and psychological tolerance. Additionally, for individuals with metal implants, such as pacemakers, MRI may pose specific safety concerns. Despite these limitations, MRI remains a crucial tool in contemporary medicine. The high-resolution images it generates, rich in structural detail, are invaluable for both scientific research and clinical practice. They help clinicians make more precise diagnoses and treatment decisions, thereby advancing medical care. As MRI technology continues to evolve, its role in promoting human health is expected to become even more significant.

1.1.1 Physics of MR

MRI is based on the fundamental physical principles of nuclear magnetic resonance (NMR). This principle was first independently established by Purcell et al. [10] and Bloch [11]. They discovered that certain atomic particles exhibit a property known as spin, which is crucial for NMR. It is important to note that not all atomic particles possess spin; specifically, only atomic nuclei with an odd number of neutrons or protons exhibit this characteristic [12]. Among the isotopes relevant to NMR imaging, such as hydrogen, sodium, and phosphorus, these elements are found in biological tissues with nearly complete abundance and possess spin. However, the concentration of these nuclei within tissues varies significantly. Hydrogen nuclei (1H) are present at ~88 mol, while sodium (23Na) and phosphorus (31P) nuclei are found at millimolar concentrations. Although sodium and phosphorus nuclei can be utilized for NMR imaging, the duration required to achieve satisfactory image quality is considerably longer compared to imaging with hydrogen nuclei. Given that hydrogen nuclei are abundant in the body, primarily in the form of water, and are present at higher concentrations, they facilitate the shortest NMR imaging times. Consequently, the majority of NMR imaging procedures predominantly rely on hydrogen nuclei.

Hydrogen nuclei possess a spin property that creates a magnetic moment, which interacts with an external magnetic field (B_0 field). When subjected to this external field, the hydrogen nuclei shift from their random spatial arrangement to align either parallel or anti-parallel to the B_0 field. They do not remain still; instead, they rotate around their alignment direction, akin to the Earth's rotation on its axis. This motion is referred to as precession. The main direction of precession aligns with the B_0 field, and the frequency of this precession is influenced by the strength of the magnetic field and the gyromagnetic ratio, represented by the equation $\omega_0 = \gamma B_0$. In this equation, ω_0 is known as the Larmor frequency, and γ denotes the gyromagnetic ratio. The precession of hydrogen nuclei around their axis produces a magnetic field that is significantly weaker than the B_0 field. When a radiofrequency (RF) pulse (B_1 field) is applied for a specific duration, and its frequency coincides with the precession frequency, it can modify the tissue's magnetization. This results in a shift of the net magnetization (magnetic moment) M from the B_0 field direction to the transverse region of the xy plane, as illustrated in Fig. 1.1b. Only the transverse component of the net magnetization, M_{xy}, is detectable by the magnetic resonance receiver. The flip angle of the net magnetization is determined by the power and duration of the RF pulse. Consequently, utilizing a 90° RF pulse for excitation can yield the highest signal strength in the transverse plane.

After the radiofrequency (RF) pulse is turned off, we notice that the transverse magnetization component, M_{xy}, slowly diminishes, while the longitudinal magnetization component, M_z, keeps rising. The decline of the nuclear magnetic resonance

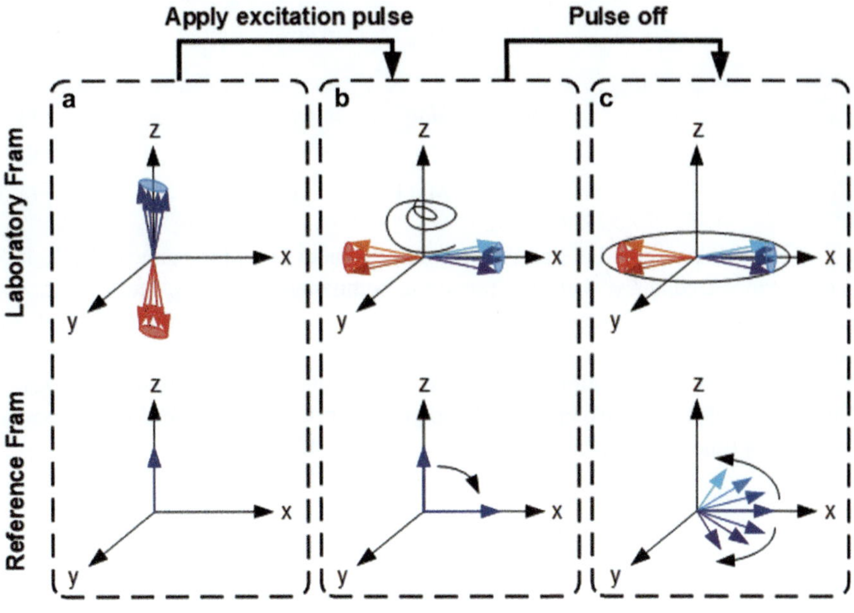

Fig. 1.1 Illustrates the rotating magnetic moment in both the stationary reference frame and the laboratory's rotating reference frame. **a** Depicts the magnetic moment in a static magnetic field B_0, **b** the magnetic moment following the application of a 90° radiofrequency pulse, and **c** represents the magnetic moment after the radiofrequency pulse has been switched off. In comparison to the spin protons that are aligned anti-parallel to the B_0 field, there are slightly more spin protons aligned parallel to the B_0 field, resulting in a net magnetization vector as seen in (**a**). Once the RF pulse is turned off, due to phase incoherence, some spins rotate at a frequency that is faster than the resonance frequency [indicated by yellow arrows in (**c**)], while others rotate at a slower frequency [shown by green arrows in (**c**)]

signal is due to the close interactions between the individual spins that constitute the magnetic moment M and their surroundings. The magnetic moment M does not continue to rotate indefinitely around the B_0 field; instead, it gradually returns to its original equilibrium state, which is aligned parallel to the B_0 field and is referred to as the equilibrium magnetization M_0. For this to happen, the spin system must release a certain amount of energy to the environment, a process known as relaxation. During relaxation, the transverse magnetization component M_{xy} processes around the B_0 field at a constant Larmor frequency, and the energy absorbed by the nuclei is transformed into a voltage. This voltage can be detected and amplified by a precisely tuned receiving coil, ultimately appearing as a free induction decay (FID) signal. To clarify how the magnetic moment returns to thermal equilibrium after the nuclear magnetic resonance signal is produced, Bloch introduced two relaxation time constants, T1 and T2. T1 indicates the rate at which the longitudinal magnetization M_z increases, while T2 represents the rate at which the transverse magnetization M_{xy} decreases.

Bloch wrote the magnetization vectors that undergo T_1 and T_2 relaxation effects as vector equations [11, 13]. Specifically, the equation of motion for the magnetic moment M when equilibrium is restored after a 90° pulse is:

$$M_x(t) = M_0 e^{-t/T_2} \sin \omega t \tag{1.1}$$

$$M_y(t) = M_0 e^{-t/T_2} \cos \omega t \tag{1.2}$$

$$M_z(t) = M_0(1 - e^{-t/T_1}) \tag{1.3}$$

These equations predict that the magnetic moment M will spiral near the B_0 field at a Larmor frequency, with the transverse component gradually decaying to zero and the longitudinal component re-growing to its original maximum value of M_0. Because T_2 is always less than T_1, the transverse magnetization component decays to zero before the longitudinal component fully recovers. The longitudinal relaxation time T_1 denotes the time it takes for M_z to grow from 0 to $(1 - 1/e)M_0$. The transverse relaxation time T_2 represents the time required for M_{xy} to decay to $(1/e)M_0$.

1.1.2 MR Image Formation

Spatial encoding in MRI requires the analysis of both frequency and phase, which can be accomplished using Fourier methods [14]. Figure 1.2 illustrates how a two-dimensional (2D) Fourier transform decomposes a brain image into 2D plane waves of varying frequencies and amplitudes. Unlike one-dimensional (1D) linear waves, these 2D plane waves can propagate in multiple directions, such as horizontally or vertically. To construct a complete image, the Fourier summation must include 2D plane waves from all directions. Individual plane waves can be thought of as layers of different sizes that stack in various orientations to form the final image.

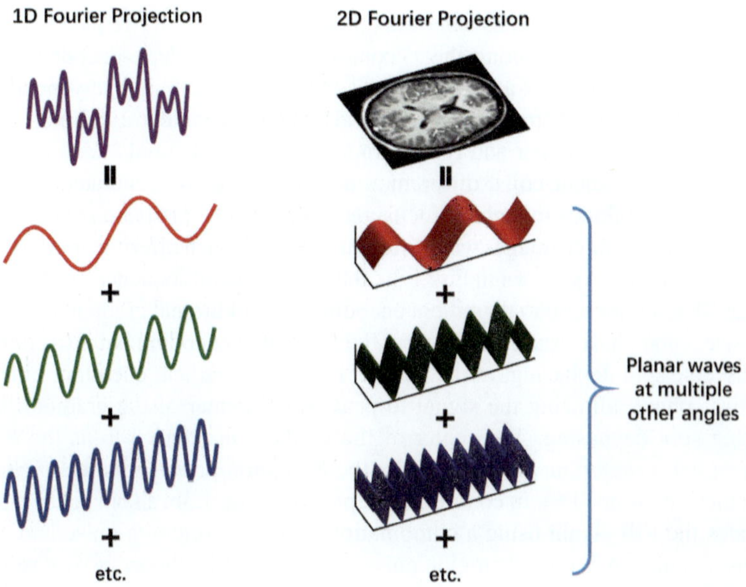

Fig. 1.2 Schematic of the Fourier expansion of a magnetic resonance image

To acquire MR signals for all plane waves, gradients are employed to introduce additional spatial linear variations in the B_0 field. Within the MRI system, gradients can be applied in any orthogonal direction using gradient coils *Gx* and *Gy*. Since the speed of the feed corresponds to the height of the MR signal, the frequency and phase can distinguish MR signals at different spatial locations.

The magnetic resonance signal at any position (x, y) within the 2D dimension can be expressed as [15]:

$$S(t) = \int_x \int_y m(x, y)e^{-i2\pi[k_x(t)x+k_y(t)y]}\mathrm{d}x\mathrm{d}y \tag{1.4}$$

which

$$K_x(t) = \frac{\gamma}{2\pi}\int_0^t G_x(\tau)\mathrm{d}\tau \tag{1.5}$$

$$K_y(t) = \frac{\gamma}{2\pi}\int_0^t G_y(\tau)\mathrm{d}\tau \tag{1.6}$$

where $m(x, y)$ is the transverse magnetization vector at position (x, y), which is related to the proton density of the imaged object at that position, the longitudinal relaxation time T_1, and the transverse relaxation time T_2; G_x is the frequency-encoded gradient strength applied along the x-axis, and G_y is the phase-encoded gradient strength applied along the y-axis; and $K_x(t)$ and $K_y(t)$ are the frequency-encoded and phase-encoded gradient strengths, respectively, versus the integral of time.

By adjusting the number and type of radiofrequency (RF) pulses, as well as the strength of the gradient coils, different echo signals can be generated. There are two primary techniques for echo refocusing: gradient echo [16] and spin echo [17]. These methods produce images with different contrasts, each offering unique clinical diagnostic benefits. As shown in Fig. 1.3a, the gradient echo sequence creates echoes through RF excitation, spatial gradient encoding, and additional gradient refocusing. Slice-select and phase gradients are essential for spatial encoding but also contribute to transverse spin dephasing. In the readout direction, gradient encoding first alters the spins' phase, allowing the signal to peak at the center of the gradient, before initiating spin dephasing. The center of the gradient is referred to as the k-space center or echo center, and the time from the excitation pulse to the echo center is called the echo time (TE). In contrast, as depicted in Fig. 1.3b, a spin echo sequence generates the MR signal using a combination of a 90° excitation pulse and a 180° refocusing pulse. After the excitation pulse, the spins gradually decay over time, and the 180° RF pulse at time τ reverses the spins' phase in the transverse plane. At time 2τ, the spins rephase along the original axis, producing the echo signal. Since the 180° RF pulse eliminates the effects of local field inhomogeneities, the spin echo sequence is primarily sensitive to T2 relaxation and is less affected by artifacts from local field variations. While spin echo provides excellent tissue contrast, it consumes more RF power than gradient echo sequences and requires a longer TE.

Currently, T1-weighted and T2-weighted images are the most commonly used scan types in hospitals. Their high contrast and resolution make them valuable for examining the brain's structural features [18–21]. These two imaging types have different sensitivities to the biophysical properties of brain tissue. T1-weighted images primarily reflect the tendency of tissues to absorb energy into their lattices, while T2-weighted images emphasize the effects of spin–spin interactions. As a result, structural changes in the brain due to diseases such as edema, inflammation,

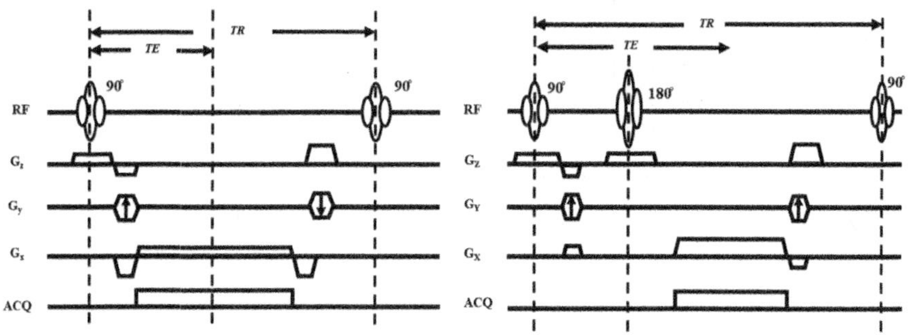

Fig. 1.3 Schematic diagram of the magnetic resonance echo sequence framework. Gradient echo sequence (**a**) and spin echo sequence (**b**) diagrams, respectively

tumors, or atrophy will appear differently in T1 and T2-weighted images. However, both T1 and T2-weighted images have a common limitation: they can only capture the macroscopic surface of tissues and cannot reveal microscopic structural details. Therefore, there is an urgent need for techniques that can provide insights into the microstructural characteristics of tissues within the body, beyond the anatomical information obtained from conventional MRI T1 and T2 imaging. Such technologies would serve as tools to unravel the complexity and diversity of the human body, enhancing our understanding of human diseases.

1.2 Basics of Diffusion MRI

Over the past two decades, diffusion magnetic resonance imaging (dMRI) has emerged as one of the most promising techniques for studying tissue microstructure, particularly the structure of brain white matter. This is due to dMRI's unique sensitivity to the movement of water molecules at the micron level. Cellular structures can impede the movement of water molecules at this scale, allowing dMRI to non-invasively reveal cellular-scale structural information within tissues [22]. This capability makes dMRI a powerful tool for probing the internal structure of the human brain and enhancing our understanding of related diseases. The advanced imaging technique helps explore the complexity of the human brain at a more microscopic level, offering new perspectives and insights into a wide range of health issues.

1.2.1 Physics of Diffusion

Diffusion is the process by which liquid molecules move randomly and chaotically, influenced by frequent collisions between molecules. This randomness is a fundamental characteristic of Brownian motion. In an ideal scenario, such as in an infinitely normalized fluid, the average distance that molecules can travel over a specific time is determined solely by their speed, which is closely related to temperature and molecular density. However, interactions with other molecular clusters can alter their movement direction. Thus, diffusion illustrates how these collisions restrict the unrestricted movement of molecules, preventing them from moving freely through space [23].

The most basic model of diffusion assumes that water molecules experience free or Gaussian diffusion. In this model, the displacement r of any molecule from its starting point is solely dependent on the time interval t. During this time, the displacement can be described by the Einstein equation [24]:

$$r^2 = 2Dt \tag{1.7}$$

where D is the diffusion coefficient, a property of spin in its diffusion medium. t is the time of diffusion motion.

Water molecules are distributed in intra- and extracellular spaces at a ratio of approximately 3:1, a balance maintained by the energy-dependent Na–K ion pump. Under certain conditions, such as poisoning or hypoxic-ischemic injury, the Na–K ion pump may fail to function properly, leading to cell swelling or rupture, significantly affecting the distribution and diffusion rate of water molecules between intra- and extracellular spaces [25]. Typically, water molecules inside the cell experience more collisions with cell walls, organelles, and macromolecules, making their diffusion more restricted than that of extracellular water molecules. Extracellular water molecules can diffuse freely, similar to their behavior in pure water, with no directional preference—a phenomenon known as isotropy, characterized by a single diffusion coefficient (D). However, due to the highly structured nature of biological tissues, the diffusion of water molecules moving through the narrow spaces between densely packed cells is quite complex. Consequently, the diffusion process in biological tissues is often restricted, resulting in a non-Gaussian distribution of diffusion distances, with different diffusion coefficients in different directions. This phenomenon is known as anisotropy. Most biological tissues exhibit some degree of anisotropy, particularly in brain white matter. In such anisotropic tissues, a single value cannot accurately describe the diffusion process. Therefore, a diffusion tensor, a 3×3 matrix, is used to describe diffusion more precisely [26]. The three diagonal elements of the diffusion tensor (D_{xx}, D_{yy}, and D_{zz}) represent the diffusion coefficients measured in each principal direction (x, y, and z) of the laboratory reference frame, while the six off-diagonal elements (D_{xy}, D_{yz}, etc.) reflect the correlations between random motions in each pair of principal directions.

$$D = \begin{bmatrix} Dxx\ Dxy\ Dxz \\ Dyx\ Dyy\ Dyz \\ Dzx\ Dzy\ Dzz \end{bmatrix} \tag{1.8}$$

For the special case of isotropic diffusion of water molecules in an isotropic organization, the non-diagonal elements are all zero, and the diagonal elements are all the same and equal to a single diffusion coefficient D (i.e. $D_{xx} = D_{yy} = D_{zz} = D$).

1.2.2 *Diffusion encoding in MRI*

dMRI is currently the only technique that can non-invasively gather information about the diffusion of water molecules in living tissues at the micron level. It measures the actual movement of water molecules by incorporating diffusion-sensitizing gradients into the magnetic resonance sequence. This allows for the recording of signal changes caused by water molecule diffusion, enabling the analysis of this motion. The resulting images display diffusion-weighted characteristics, known as Diffusion-Weighted Imaging (DWI).

DWI is produced by applying gradient pulses at both ends of a spin-echo imaging sequence. For quantitative analysis, it is essential to accurately assess the level of diffusion weighting in the sequence. In dMRI, the "*b*-value" is commonly used to quantify the extent of diffusion weighting:

$$b = \gamma^2 \int_0^{\mathrm{TE}} \left[\int_0^{t} g(t')\mathrm{d}t' \right] \mathrm{d}t \tag{1.9}$$

where $g(t')$ is the gradient pulse as a function of time. *b*-values are defined in such a way as to give a simplified expression of the signal amplitude decay:

$$S = S_0 e^{(-bD)} \tag{1.10}$$

where S is the signal intensity obtained for each voxel within the image after applying a diffusion gradient, S_0 is the corresponding signal intensity for each voxel without applying a diffusion gradient, i.e. $b = 0$.

The diffusion coefficient D for each direction of the diffusion gradient can be derived from the associated diffusion-weighted images (DWI). Tanner introduced the concept of the Apparent Diffusion Coefficient [27] (ADC) to account for the combined effects of various complex, incoherent movements of water molecules. Essentially, ADC indicates the average diffusion level of water molecules across all diffusion directions within a specific voxel of the image. By calculating the ADC for each voxel in a DWI, an ADC map is generated, as illustrated in Fig. 1.4. This type of image effectively showcases the apparent diffusion characteristics and is highly valuable in numerous medical applications, such as mapping the fiber pathways of brain white matter and assessing tumor treatment responses. The detailed insights into water molecule diffusion at the micron scale in biological tissues offer a unique viewpoint for investigating the microscopic environment within the body.

While ADC provides general information regarding the diffusion of water molecules, it has limitations in capturing the complexity of diffusion patterns. To better understand the diffusion patterns of water molecules both inside and outside cells in various directions, more sophisticated diffusion tensor imaging [26] (DTI) technology is necessary. DTI calculates the diffusion tensor, which provides insights into anisotropic diffusion. This tensor can be utilized to determine several key diffusion metrics, including fractional anisotropy (FA), mean diffusivity (MD), axial diffusivity (AD), and radial diffusivity (RD). DTI is a straightforward and effective method that has been extensively used to investigate different aspects of biological tissues, especially the structure of white matter in the brain. In the white matter, the presence of axons and myelin leads to water diffusion primarily occurring along the direction of nerve fiber bundles. DTI can map the orientation of white matter fiber tracts and,

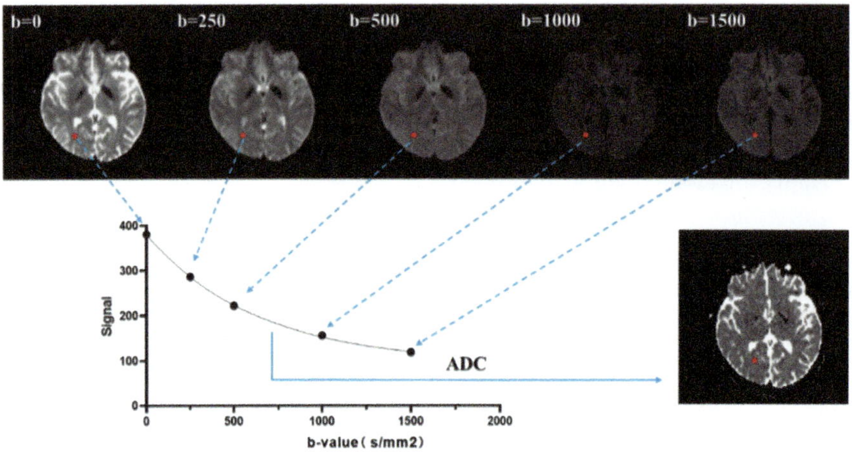

Fig. 1.4 Diffusion-weighted image with ADC image. Adapted from Dan Wu (2024). Principles and applications of magnetic resonance imaging. Science Press

through methods like fiber-tracking [28] or tractography [29], can non-invasively examine the brain's connectivity.

However, the main drawback of Diffusion Tensor Imaging (DTI) is that it represents the diffusion tensor with just six parameters, resulting in a simplified model. It presumes that all axons within a voxel are aligned in the same direction. Since voxels in the brain usually contain axons oriented in various directions, this assumption is often inaccurate for many brain areas. To address the limitations of DTI and effectively map the intricate orientation of axonal fiber bundles within each voxel, several more sophisticated diffusion-based techniques have been developed. These methods require longer acquisition times and do not depend on the tensor model, thus bypassing the limitations of the simplistic six-parameter tensor model. Examples of these techniques include diffusion spectrum imaging [30] (DSI), q-ball imaging [31] (QBI), and high-angular resolution diffusion imaging [32] (HARDI). They provide more comprehensive information for each voxel, offering greater insights into fiber tracts.

1.3 Advanced Diffusion MRI

Microstructure modeling is an advanced imaging method that has developed from traditional dMRI. This approach systematically probes water diffusion behaviors in biological microenvironments by varying several diffusion-weighted parameters, such as gradient strength and direction, as well as diffusion time. It then uses mathematical and physical models to derive important microstructural characteristics and uncover tissue organization at the cellular level. Two main strategies have emerged: *q*-space microstructure modeling, which manipulates diffusion gradient parameters [33], and *t*-space microstructure modeling, which alters diffusion time characteristics [34]. The *t*-space paradigm uses geometrically simplified compartments to represent tissue microstructure. By establishing theoretical signal expressions based on fundamental physical principles under constrained assumptions, this framework enables inverse estimation of microstructural properties from experimentally measured signals. For instance, cancer cells can be modeled as closely-packed spheres while white matter fibers are abstracted as cylindrical structures with varying orientations. These representations not only provide intuitive visualizations but also establish specific correlations with biophysical parameters including cellular dimensions, intracellular water fraction, membrane surface-to-volume ratio, and permeability coefficients [35–38].

1.3.1 Q-space Diffusion MRI

Within the dMRI framework, DWI is a fundamental model rooted in the core assumption that water molecules undergo isotropic Gaussian diffusion in a homogeneous medium. However, the applicability of this assumption is limited in highly structured biological tissues, such as cerebral white matter, where the motion of water molecules is significantly restricted by structures like cell membranes and axons, leading to pronounced anisotropy. To more accurately characterize this complex diffusion behavior, DTI [11] was developed. DTI no longer simplifies diffusion to a single scalar value but employs a third-order tensor (a 3x3 symmetric matrix) to comprehensively describe the three-dimensional diffusion ellipsoid of water molecules. Its key parameters include Mean/Axial/Radial Diffusivity (MD/AD/RD), which reflect the overall diffusion efficiency, and Fractional Anisotropy (FA), which quantifies the directional coherence of diffusion. The FA value is positively correlated with the structural integrity and directionality of the tissue; a reduction in its value typically

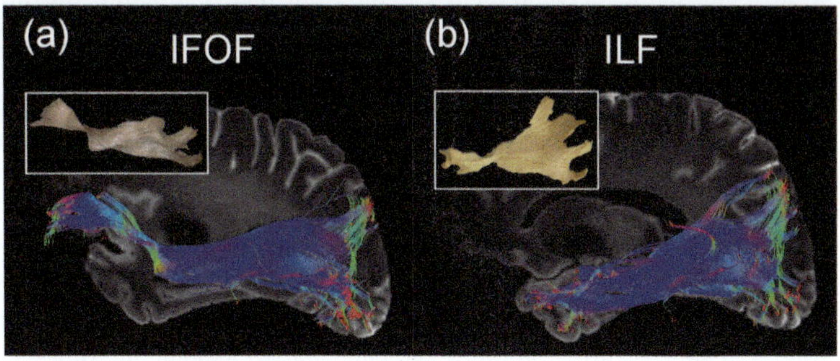

Fig. 1.5 dMRI white matter fiber tracking. **a** Inferior fronto-occipital fasciculus (IFOF) fiber and **b** inferior longitudinal fasciculus (ILF) fiber

indicates fiber tract damage or demyelination. To further resolve the more complex microenvironments composed of organelles and membrane structures, the Diffusion Kurtosis Imaging (DKI) [12] model was proposed. This model more effectively reflects the microstructural complexity and heterogeneity of tissue by quantifying the degree to which water diffusion deviates from an ideal Gaussian distribution. Its core parameters, Mean/Axial/Radial Kurtosis (MK/AK/RK), are positively correlated with the inhomogeneity and restrictedness of the diffusion environment, thus indirectly reflecting cell density or the complexity of tissue architecture.

Based on these models, one of the most valuable applications of dMRI lies in the use of fiber tracking (or tractography) techniques [39] to reconstruct the major white matter pathways in the brain, and subsequently, to infer the macroscopic neural connectivity map, known as the "connectome." This technique offers an unprecedented, non-invasive window for exploring the complex structural wiring of the brain and its underlying principles, and it supports research into the relationships between brain structure and cognitive, behavioral, and genetic factors. Furthermore, advancements in 3D MRI sequences and hardware systems have now enabled whole-brain tractography at higher spatial resolutions. This allows researchers to identify and reconstruct fine structures that are unobservable in conventional low-resolution scans, such as the short-range U-fibers connecting adjacent gyri, thereby greatly enhancing the fineness and accuracy of brain connectivity maps (as shown in Figure 1.5) [40].

Consequently, dMRI plays a pivotal role in major scientific initiatives aimed at mapping the human brain on a large scale, including the Human Connectome Project (HCP) [41], the Adolescent Brain Cognitive Development (ABCD) study [42], and the UK Biobank project [43].

1.3.2 Time-dependent Diffusion MRI

Initial investigations commonly posited that water molecules within the white matter of the human brain exhibited no diffusion time-dependency [44]. This postulation might be attributed to the fact that, relative to the white matter structure, the diffusion time at that juncture was already in the long-time domain, thereby leading to diminished sensitivity to the microstructure of white matter. With the advancements in hardware performance and the innovation in diffusion sequence design, the measurement scope of diffusion time has been substantially extended. Among these developments, the time—dependent diffusion magnetic resonance imaging (TDDMRI) acquisition technique is particularly notable. This technique encompasses the novel oscillating gradient spin echo (OGSE) [45] diffusion—weighted imaging technique and the conventional pulsed gradient spin echo (PGSE) [46] diffusion—weighted imaging technique. In comparison to PGSE, the OGSE technique is capable of acquiring dMRI signals at shorter diffusion times and can clearly elucidate the time—dependency of diffusion signals in the brain. Besides the direct utilization of OG-dMRI signals with short diffusion times, the application of the OGSE technique to detect the time—dependency of DTI metrics (such as ADC) also holds significant scientific value. The ADC measured by DTI is contingent upon the time consumed by water molecules to explore the local environment, namely the effective diffusion time (Δ_{eff}). As the diffusion time lengthens, water molecules will interact with an increasing number of obstacles, and the ADC will consequently decline, ultimately approaching an asymptotic low value. Hence, by measuring the variation law of ADC with respect to diffusion time, a more profound comprehension of the microstructure within the tissue can be attained.

Xu et al. [47] formulated the IMPULSED (imaging microstructural parameters using limited spectrally edited diffusion) model grounded on the diffusion time dependence within tumor tissues, aiming to simulate microstructural parameters such as the diameter and volume fraction of cancer cells. In the case of structures featuring negligible water exchange between the intracellular and extracellular spaces, the signal under long time domain diffusion would, to a certain degree, mirror the overall structural information of the tissue. The variation of this signal with respect to diffusion time is contingent upon the power-law relationship inherent in the tissue structure. Conversely, within a multi-compartment model characterized by water exchange between the intracellular and extracellular spaces, the water exchange among compartments disrupts this power-law relationship, thereby resulting in diverse forms of diffusion time dependence of the diffusion parameters. In practical scenarios, it is highly probable that these two mechanisms coexist. Novikov et al. [48] postulated that for weakly heterogeneous media in the long time domain, diffusion admits a perturbative solution. Based on this premise, they deduced the power-law equation representation of diffusion over time. The expressions of the diffusion coefficient in the conventional sense can be denoted by Eqs. 1.11 and 1.12:

$$D(t) = D_\infty + B \cdot t^{-\vartheta} \tag{1.11}$$

$$\vartheta = \frac{p+d}{2} \tag{1.12}$$

Here, ϑ represents the diffusion (dynamic) exponent, p denotes the global structural complexity, and d is the spatial dimension. Structures with higher complexity (smaller p values) exhibit smaller dynamic exponents. Since different p and d values can correspond to specific structural barriers, it is theoretically possible to infer the global structural characteristics of tissues by observing the time-dependent features of diffusion signals. Figure 1.6 illustrates the specific scenarios corresponding to different tissue structures with varying values.

OGSE, as the preeminent technique within the framework of TDDMRI, exhibits the capacity to discern shorter diffusion time scales in contrast to the conventional PGSE. The integration of the OGSE and PGSE techniques affords the opportunity to observe the diffusion time dependence of microstructural metrics across diverse biological tissues [45, 49–53], encompassing the cortical gray matter and hippocampus of mice, as well as the principal white matter tracts and deep gray matter regions of the human brain. A recent ex-vivo investigation on the murine brain

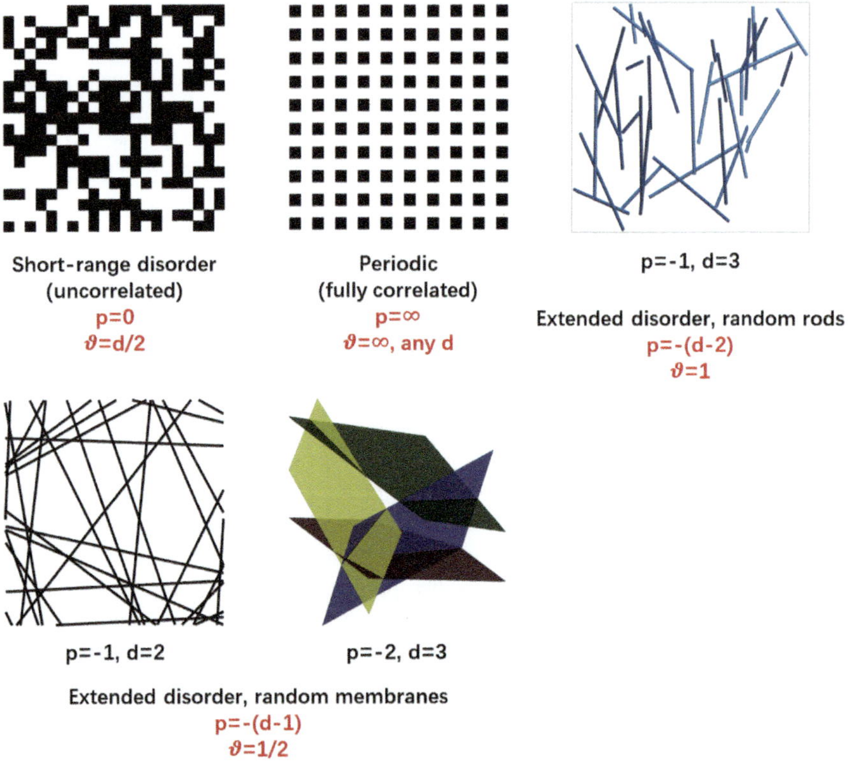

Fig. 1.6 Dynamic coefficients corresponding to different structures. And the fluctuation function of the extended disordered structure

demonstrated that under conditions of shorter diffusion times, the tissue contrast among distinct cell types in the hippocampus of the murine brain was markedly enhanced [54]. This study further revealed that as the diffusion time diminished, the eigenvalues of the diffusion tensor (both parallel and perpendicular) increased, while the FA within each white matter and gray matter region decreased. Analogous findings were also observed in the white matter and gray matter of the in-vivo rat brain [55], signifying that alterations in the diffusion time engendered changes in the diffusion parameters. These studies have ascertained that in the majority of brain regions, ADC displays diffusion time dependence, manifesting as an increase with the elongation of the diffusion time. This diffusion time dependence uncovers microstructural information that remains undetected by the PGSE technique alone. For instance, this diffusion time dependence implies the presence of diffusion restrictions within the lamellar keratin layers of cysts, a phenomenon that cannot be elucidated solely by the PGSE technique [56].

TDDMRI holds the capacity to delve more profoundly into the microstructural features of tumor cells and inherently possesses an advantage in the detection of tumor cell density. It has emerged as one of the pivotal clinical tools for the evaluation, classification, and prediction of tumors, furnishing an effective approach for the diagnosis of brain tumors. In this context, researchers have discerned that shorter diffusion times are capable of yielding higher diffusion signal contrast or ADC disparities, and are anticipated to proffer a non-invasive methodology for differentiating between benign and malignant tumors, diverse tumor types, as well as tumors and normal tissues [57–62]. In multiple investigations concerning mouse tumor models, it was ascertained that there was no discernible difference in the ADC of tumors and normal tissues when measured by the PGSE technique. Conversely, the ADC of tumors measured by the OGSE technique was found to be higher than that of normal tissues. In the realm of clinical practice, recent research endeavors are also exploring the feasibility and efficacy of implementing the OGSE technique in clinical systems [47, 56, 63–67]. In a particular study, Xu et al. [47] demonstrated that on a 3 T clinical scanner, by means of the TDDMRI technique employing simplified PGSE and OGSE, in conjunction with an advanced biophysical model, the cell size in breast cancer could be accurately estimated. Tetreault et al. [68] utilized the TDDMRI technique of OGSE and PGSE to measure the time dependence of diffusion parameters in the human corpus callosum with respect to age and gender variations. Wu et al. [69] discovered that employing TDDMRI with OGSE and PGSE techniques to measure ADC values at different diffusion times enabled more accurate characterization of different cell types, thereby rendering the classification of tumors more precise.

1.4 Objective and Organization of This Book

This book focuses on optimizing time-dependent diffusion MRI (TDDMRI) imaging in several aspects, including reducing scan time, improving image resolution and signal-to-noise ratio (SNR), correcting phase errors between multi-segment acquisitions, and eliminating cerebrospinal fluid (CSF) effects. This paper first replaces the 2D EPI readout module in the conventional TDDMRI sequence with a 3D GRASE readout module and employs 3D GRAPPA (generalized automatic calibration part parallel acquisition) parallel imaging acceleration to reduce echo time (TE) and repetition time (TR), thereby enhancing image resolution, SNR, and scan speed. Subsequently, a 3D navigator module is added after the readout module to correct phase errors between multi-segment acquisitions, resulting in high-resolution whole-brain TDDMRI images. Finally, an inversion recovery (IR) module is used to eliminate the influence of CSF signals on surrounding tissues. The paper is divided into seven chapters, with the content of each chapter summarized as follows:

Chapter 1: Introduction: This chapter first introduces the fundamental principles of MRI technology to the readers. Based on this, it further elaborates on the basic principles and clinical applications of dMRI and its extension, TDDMRI. Subsequently, it identifies the key research issues that remain to be addressed in current TDDMRI and summarizes the relevant progress in this field. Finally, the main framework of the book is presented.

Chapter 2: Basics of Diffusion MRI Sequence: This chapter provides an in-depth analysis of dMRI sequence design. It first offers a detailed introduction to the design of dMRI imaging sequences, specifically elaborating on various techniques in the diffusion preparation module and the signal readout module. Subsequently, it provides a comprehensive description of high-resolution and high-SNR dMRI imaging techniques. Finally, it systematically discusses fast imaging techniques, aiming to shorten scan time and improve the efficiency of dMRI imaging. This chapter is rich in content and theoretical knowledge, and it is hoped that it will help readers gain a deeper understanding of dMRI imaging techniques, providing theoretical support for the specific technical developments in the following chapters.

Chapter 3: 3D Diffusion Gradient Prepared Gradient and Spin Echo Imaging: This chapter proposes a series of innovative solutions to address the issues of long scan time, low image resolution, and low SNR in TDDMRI imaging at short diffusion times. First, it replaces the conventional 2D EPI readout module in the TDDMRI sequence with a 3D GRASE readout module. This improvement significantly reduces TE and TR, effectively enhancing spatial resolution and SNR. Second, it achieves separation of diffusion encoding and readout encoding in the conventional TDDMRI sequence. Finally, through a series of experiments, it demonstrates that compared to the conventional 2D EPI readout, the 3D GRASE readout can effectively improve the SNR of TDDMRI images under the same acquisition time.

Chapter 4: Twin-Navigator based 3D Gradient and Spin Echo Imaging: This chapter proposes a TDDMRI imaging sequence based on dual-navigator technology for multi-segment acquisition. This technique effectively corrects inter-segment phase errors in multi-segment acquisitions, eliminates motion-induced aliasing artifacts, and achieves high-fidelity whole-brain imaging. Second, the addition of 2D GRAPPA acceleration in the EPI phase-encoding direction of the GRASE readout module shortens TE and improves image SNR. Compared to conventional 2D sequences, this sequence not only reduces scan time under the same SNR but also provides more accurate results in fiber tracking and microstructural parameter estimation.

Chapter 5: 3D Navigator based 3D Gradient Spin Echo Imaging: This chapter further optimizes the dual-navigator TDDMRI imaging sequence presented in Chap. 4 by incorporating 3D navigator technology. Compared to dual-navigator technology, 3D navigator technology can more effectively correct signal phase errors and eliminate motion artifacts in multi-segment acquisitions. In addition, this chapter implements 3D GRAPPA acceleration technology based on 2D GRAPPA, achieving acceleration in both phase-encoding directions of the GRASE readout module.

Chapter 6: IR-based 3D Gradient and Spin Echo Imaging: This chapter focuses on optimizing TDDMRI imaging in regions near CSF by replacing the global stabilization module with an inversion recovery (IR) module. This change effectively suppresses CSF effects and improves the accuracy of measuring ADC diffusion time dependence. This optimized technique was experimentally validated on an internationally advanced Siemens Prisma 3 T MRI scanner. The results demonstrate that this technique also plays an important role in accurately measuring the diffusion time dependence of ADC in tumor regions. This finding has profound implications for future clinical diagnosis and treatment of tumors.

References

1. Plewes DB, Kucharczyk W (2012) Physics of MRI: a primer. J Magnet Reson Imag 35(5):1038–1054
2. Hashemi RH, Bradley WG, Lisanti CJ (2012) MRI: the basics. Lippincott Williams & Wilkins
3. Katti G, Ara SA, Shireen A (2011) Magnetic resonance imaging (MRI): a review. Int J Dental Clin 3(1):65–70
4. Warren BE (1990) X-ray diffraction. Courier Corporation
5. Buzug TM (2011) Computed tomography, Springer handbook of medical technology. Springer, New York, pp 311–342
6. Buxton RB, Edelman RR, Rosen BR et al (1987) Contrast in rapid MR imaging: T1- and T2-weighted imaging. J Comput Assist Tomogr 11(1):7–16
7. Abdel-Aty H, Simonetti O, Friedrich MG (2007) T2-weighted cardiovascular magnetic resonance imaging. J Magnet Reson Imag Off J Int Soc Magnet Reson Med 26(3):452–459
8. Baliyan V, Das CJ, Sharma R et al (2016) Diffusion weighted imaging: technique and applications. World J Radiol 8(9):785
9. Dale BM, Brown MA, Semelka RC (2015) MRI: basic principles and applications. Wiley, Amsterdam
10. Purcell EM, Torrey HC, Pound RV (1946) Resonance absorption by nuclear magnetic moments in a solid. Phys Rev 69(1–2):37
11. Bloch F (1946) Nuclear induction. Phys Rev 70(7–8):460
12. Atta-Ur-Rahman T (2012) Nuclear magnetic resonance: basic principles. Springer, New York
13. Bloch F, Hansen WW, Packard M (1946) The nuclear induction experiment. Phys Rev 70(7–8):474–485
14. Bracewell RN (1989) The fourier transform. Sci Am 260(6):86–95
15. Nishimura DG (1996) Principles of magnetic resonance imaging. Stanford University
16. Markl M, Leupold J (2012) Gradient echo imaging. J Magnet Reson Imag 35(6):1274–1289
17. Jung BA, Weigel M (2013) Spin echo magnetic resonance imaging. J Magnet Reson Imag 37(4):805–817
18. Raz E, Cercignani M, Sbardella E et al (2010) Gray-and white-matter changes 1 year after first clinical episode of multiple sclerosis: MR imaging. Radiology 257(2):448–454
19. Welker KM, Patton A (2012) Assessment of normal myelination with magnetic resonance imaging. Semin Neurol 2012:015–028
20. Zhou Y, Kierans A, Kenul D et al (2013) Mild traumatic brain injury: longitudinal regional brain volume changes. Radiology 267(3):880–890
21. Steenweg ME, Vanderver A, Blaser S et al (2010) Magnetic resonance imaging pattern recognition in hypomyelinating disorders. Brain 133(10):2971–2982
22. Tournier J-D (2019) Diffusion MRI in the brain: theory and concepts. Prog Nucl Magnet Reson Spectrosc 112:1–16
23. Gore JC, Xu J, Colvin DC et al (2010) Characterization of tissue structure at varying length scales using temporal diffusion spectroscopy. NMR Biomed 23(7):745–756
24. Einstein A (1956) Investigations on the theory of the brownian movement. Courier Corporation
25. Minati L, Węglarz WP (2007) Physical foundations, models, and methods of diffusion magnetic resonance imaging of the brain: a review. Concep Magnet Reson A Educ J 30(5):278–307
26. Lope-Piedrafita S (2018) Diffusion tensor imaging (DTI). Preclin MRI Methods Protocols 34:103–116
27. Herneth AM, Guccione S, Bednarski M (2003) Apparent diffusion coefficient: a quantitative parameter for in vivo tumor characterization. Eur J Radiol 45(3):208–213

28. Morris DM, Embleton KV, Parker GJ (2008) Probabilistic fibre tracking: differentiation of connections from chance events. Neuroimage 42(4):1329–1339
29. Yamada K, Sakai K, Akazawa K et al (2009) MR tractography: a review of its clinical applications. Magnet Reson Med Sci 8(4):165–174
30. Wedeen VJ, Hagmann P, Tseng WYI et al (2005) Mapping complex tissue architecture with diffusion spectrum magnetic resonance imaging. Magnet Reson Med 54(6):1377–1386
31. Tuch DS (2004) Q-ball imaging. Magnet Reson Med Off J Int Soc Magnet Reson Med 52(6):1358–1372
32. Frank LR (2002) Characterization of anisotropy in high angular resolution diffusion-weighted MRI. Magnet Reson Med Off J Int Soc Magnet Reson Med 47(6):1083–1099
33. Basser PJ (2002) Relationships between diffusion tensor and q-space MRI. Magnet Resonan Med Off J Int Soc Magnet Reson Med 47(2):392–397
34. Novikov DS, Fieremans E, Jespersen SN et al (2019) Quantifying brain microstructure with diffusion MRI: theory and parameter estimation. NMR Biomed 32(4):e3998
35. Portnoy S, Fichtner ND, Dziegielewski C et al (2014) In vitro detection of apoptosis using oscillating and pulsed gradient diffusion magnetic resonance imaging. NMR Biomed 27(4):371–380
36. Reynaud O, Winters KV, Hoang DM et al (2016) Pulsed and oscillating gradient MRI for assessment of cell size and extracellular space (POMACE) in mouse gliomas. NMR Biomed 29(10):1350–1363
37. Jiang X, Li H, Xie J et al (2017) In vivo imaging of cancer cell size and cellularity using temporal diffusion spectroscopy. Magnet Reson Med 78(1):156–164
38. Li H, Gore JC, Xu J (2014) Fast and robust measurement of microstructural dimensions using temporal diffusion spectroscopy. J Magnet Reson 242:4–9
39. Jeurissen B, Descoteaux M, Mori S et al (2019) Diffusion MRI fiber tractography of the brain. NMR Biomed 32(4):e3785
40. Song AW, Chang H-C, Petty C, et al (2014) Improved delineation of short cortical association fibers and gray/white matter boundary using whole-brain three-dimensional diffusion tensor imaging at submillimeter spatial resolution. Brain connect 4(9):636–640
41. Van Essen DC, Ugurbil K (2012) The future of the human connectome. Neuroimage 62(2):1299–1310
42. Casey BJ, Cannonier T, Conley MI et al (2018) The adolescent brain cognitive development (ABCD) study: imaging acquisition across 21 sites. Develop Cognit Neurosci 32:43–54
43. Miller KL, Alfaro-Almagro F, Bangerter NK et al (2016) Multimodal population brain imaging in the UK Biobank prospective epidemiological study. Nat Neurosci 19(11):1523–1536
44. Clark CA, Hedehus M, Moseley ME (2001) Diffusion time dependence of the apparent diffusion tensor in healthy human brain and white matter disease. Magnet Reson Med 45(6):1126–1129
45. Baron CA, Beaulieu C (2014) Oscillating gradient spin-echo (OGSE) diffusion tensor imaging of the human brain. Magnet Reson Med 72(3):726–736
46. Stejskal EO, Tanner JE (1965) Spin diffusion measurements: spin echoes in the presence of a time-dependent field gradient. J Chem Phys 42(1):288–292
47. Xu J, Jiang X, Li H et al (2020) Magnetic resonance imaging of mean cell size in human breast tumors. Magnet Reson Med 83(6):2002–2014
48. Novikov DS, Jensen JH, Helpern JA et al (2014) Revealing mesoscopic structural universality with diffusion. Proceed Natl Acad Sci USAm 111(14):5088–5093
49. Palombo M, Ianus A, Guerreri M et al (2020) SANDI: A compartment-based model for non-invasive apparent soma and neurite imaging by diffusion MRI. Neuroimage 215:116835
50. Wu D, Zhang J (2017) The effect of microcirculatory flow on oscillating gradient diffusion MRI and diffusion encoding with dual-frequency orthogonal gradients (DEFOG). Magnet Reson Med 77(4):1583–1592
51. Pyatigorskaya N, Le Bihan D, Reynaud O et al (2014) Relationship between the diffusion time and the diffusion MRI signal observed at 17.2 tesla in the healthy rat brain cortex. Magnet Reson Med 72(2):492–500

52. Wu D, Martin LJ, Northington FJ et al (2014) Oscillating gradient diffusion MRI reveals unique microstructural information in normal and hypoxia-ischemia injured mouse brains. Magnet Reson Med 72(5):1366–1374
53. Van AT, Holdsworth SJ, Bammer R (2014) In vivo investigation of restricted diffusion in the human brain with optimized oscillating diffusion gradient encoding. Magnet Reson Med 71(1):83–94
54. Aggarwal M, Jones MV, Calabresi PA et al (2012) Probing mouse brain microstructure using oscillating gradient diffusion MRI. Magnet Reson Med 67(1):98–109
55. Kershaw J, Leuze C, Aoki I et al (2013) Systematic changes to the apparent diffusion tensor of in vivo rat brain measured with an oscillating-gradient spin-echo sequence. Neuroimage 70:10–20
56. Andica C, Hori M, Kamiya K et al (2018) Spatial restriction within intracranial epidermoid cysts observed using short diffusion-time diffusion-weighted imaging. Magnet Reson Med Sci 17(3):269–272
57. Bongers A, Hau E, Shen H (2018) Short diffusion time diffusion-weighted imaging with oscillating gradient preparation as an early magnetic resonance imaging biomarker for radiation therapy response monitoring in glioblastoma: a preclinical feasibility study. Int J Radiat Oncol Biol Phys 102(4):1014–1023
58. Colvin DC, Loveless ME, Does MD et al (2011) Earlier detection of tumor treatment response using magnetic resonance diffusion imaging with oscillating gradients. Magnet Reson Imag 29(3):315–323
59. Xu J, Li K, Smith RA et al (2012) Characterizing tumor response to chemotherapy at various length scales using temporal diffusion spectroscopy. PLoS ONE 7(7):e41714
60. Zhu A, Shih R, Huang RY et al (2023) Revealing tumor microstructure with oscillating diffusion encoding MRI in pre-surgical and post-treatment glioma patients. Magnet Reson Med 90(5):1789–1801
61. Iima M, Yamamoto A, Kataoka M et al (2019) Time-dependent diffusion MRI to distinguish malignant from benign head and neck tumors. J Magnet Reson Imag 50(1):88–95
62. Colvin DC, Yankeelov TE, Does MD et al (2008) New insights into tumor microstructure using temporal diffusion spectroscopy. Cancer Res 68(14):5941–5947
63. Baron CA, Kate M, Gioia L et al (2015) Reduction of diffusion-weighted imaging contrast of acute ischemic stroke at short diffusion times. Stroke 46(8):2136–2141
64. Boonrod A, Hagiwara A, Hori M et al (2018) Reduced visualization of cerebral infarction on diffusion-weighted images with short diffusion times. Neuroradiology 60:979–982
65. Döring A, Kreis R (2019) Magnetic resonance spectroscopy extended by oscillating diffusion gradients: cell-specific anomalous diffusion as a probe for tissue microstructure in human brain. Neuroimage 202:116075
66. Maekawa T, Hori M, Murata K et al (2019) Choroid plexus cysts analyzed using diffusion-weighted imaging with short diffusion-time. Magnet Reson Imag 57:323–327
67. Tan ET, Shih RY, Mitra J et al (2020) Oscillating diffusion-encoding with a high gradient-amplitude and high slew-rate head-only gradient for human brain imaging. Magnet Reson Med 84(2):950–965
68. Tétreault P, Harkins KD, Baron CA et al (2020) Diffusion time dependency along the human corpus callosum and exploration of age and sex differences as assessed by oscillating gradient spin-echo diffusion tensor imaging. Neuroimage 210:116533
69. Wu D, Jiang K, Li H et al (2022) Time-dependent diffusion MRI for quantitative microstructural mapping of prostate cancer. Radiology 303(3):578–587

Chapter 2
Basics of Diffusion MRI Sequence

Abstract Diffusion MRI (dMRI) is a powerful imaging technique that leverages the diffusion of water molecules within tissues to generate contrast in MRI images. This method provides unique insights into the microstructural properties of biological tissues, particularly the brain's white matter. By measuring the diffusion of water molecules, dMRI can reveal the underlying architecture of tissues at a microscopic level, which is often beyond the reach of conventional MRI techniques. The core principle of dMRI lies in the diffusion of water molecules, a process driven by thermal energy. Water molecules in biological tissues move randomly, and their movement is influenced by the microenvironment, including cell membranes, organelles, and other structural barriers. This random movement, known as Brownian motion, is characterized by frequent collisions between molecules, which constrain their mobility. In tissues with well-defined structures, such as white matter in the brain, water molecules tend to diffuse along specific directions, reflecting the underlying fiber orientation. This anisotropic diffusion is a key feature that dMRI exploits to generate detailed images of tissue microstructure.

Keywords Sequence · High resolution · Fast imaging

2.1 Compositions of Diffusion MRI Sequence

The typical dMRI sequence primarily consists of two modules: the diffusion preparation module and the signal readout module, as illustrated in Fig. 2.1. These two modules form the fundamental framework of the dMRI sequence. The diffusion preparation module incorporates various commonly used techniques, such as conventional Pulsed Gradient Spin Echo (PGSE) encoding [1] and Oscillating Gradient Spin Echo (OGSE) encoding for short diffusion times [2]. This module induces attenuation of the diffusion signal, thereby creating a specific diffusion contrast. Subsequently, the signal readout module captures the diffusion signal under this diffusion contrast. Commonly used readout module techniques include Fast Spin Echo (FSE) [3], Echo Planar Imaging (EPI) [4], and Gradient and Spin Echo Imaging (GRASE) [5]. This

D. Wu et al., *Development and Application of 3D Diffusion Magnetic Resonance Imaging Sequences*, Advanced Topics in Science and Technology in China 5,
https://doi.org/10.1007/978-981-95-0808-2_2

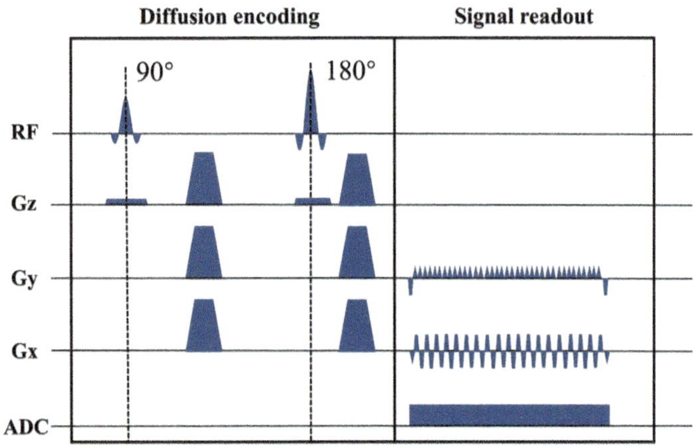

Fig. 2.1 Schematic diagram of dMRI sequence framework

framework vividly demonstrates the working principle of the dMRI sequence. It aids in a better understanding and optimization of dMRI sequences, thereby promoting the development of dMRI imaging technology.

2.1.1 Diffusion-Prepared Module

(1) PGSE

Modern dMRI sequences trace their origins back to the mid-1960s, when Stejskal and Tanner [6] developed the widely accepted and utilized Stejskal–Tanner sequence, which is now known as the Pulsed Gradient Spin Echo (PGSE) sequence, as shown in Fig. 2.2. The PGSE sequence employs a 90° radiofrequency (RF) pulse to excite the imaging region and a 180° RF pulse to refocus the signal. Symmetric strong diffusion gradients (DG) are applied on either side of the 180° RF pulse. For magnetized protons in a stationary state, their phase is not affected by the DG. This is because any phase accumulated during the first application of the DG is reversed by the second application, resulting in complete cancellation of the phase accumulation. However, for protons undergoing diffusion, the DG causes dephasing, ultimately leading to signal attenuation. The DG can be oriented in any direction or combination of directions in the coordinate system to achieve specific diffusion gradient encoding directions. The signal attenuation in the dMRI imaging process can be calculated using Eqs. 2.1 and 2.2:

$$S = S_0 e^{(-bD)} \tag{2.1}$$

$$b = \gamma^2 G^2 \delta^2 (\Delta - \delta/3) \tag{2.2}$$

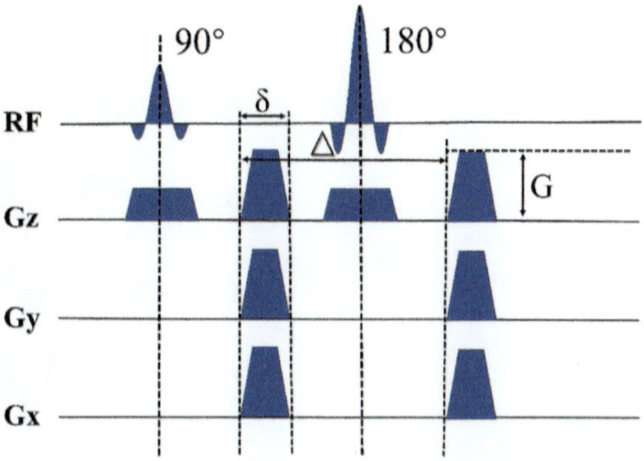

Fig. 2.2 Schematic diagram of PGSE sequence framework

where S_0 represents the baseline MRI signal intensity in the absence of the diffusion gradient (DG), while S is the signal intensity after the DG is applied. The parameter D denotes the diffusion coefficient of water molecules, and G is the strength of the diffusion gradient. The symbols δ and Δ represent the duration of the DG pulse and the time interval between the two DG pulses, respectively. The b-value, which quantifies the degree of diffusion weighting, depends on the strength, duration, and interval of the two DG pulses. Therefore, increasing the gradient amplitude, shortening the pulse duration, and lengthening the interval between the gradients can result in a higher b-value.

From a microscopic perspective, the PGSE sequence measures the average displacement of water molecules over an effective diffusion time (Δ_{eff}), which is typically defined as:

$$\Delta_{\text{eff}} = (\Delta - \delta/3) \tag{2.3}$$

(2) OGSE

When using the PGSE sequence for imaging with short diffusion times, according to Eqs. 2.2 and 2.3, reducing the diffusion time typically requires decreasing the values of Δ and δ, while increasing the gradient strength G to maintain a consistent b-value. As a result, the lower limit of diffusion time is constrained by the maximum gradient strength, which is determined by the hardware capabilities of the MRI system. In practice, achieving diffusion times below 10 ms using the PGSE sequence is nearly impossible.

To overcome this limitation and probe spin diffusion at shorter time scales, the OGSE sequence was introduced [2], as illustrated in Fig. 2.3. The OGSE sequence employs a series of short-duration diffusion-weighted gradients applied in rapid succession to achieve a low effective diffusion time (Δ_{eff}). Because each diffusion-weighted period is adjacent to the next, they cannot be considered independent. The diffusion time in OGSE can be better understood by transforming the diffusion tensor into a frequency-dependent form $D(f)$, where the signal attenuation S/S_0 is given

Fig. 2.3 Schematic diagram of OGSE sequence framework

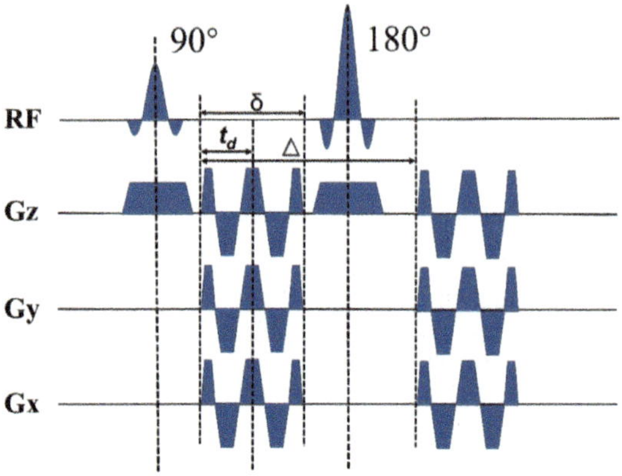

by Eq. 2.4:

$$\text{In}(S/S_0) = \int F(f)D(f)F(f)\mathrm{d}f \tag{2.4}$$

where $F(f)$ is the frequency domain representation of the cumulative effect of the diffusion gradient on the proton phase:

$$F(f) = -\int_0^\infty \left(\int_0^t \gamma G(\tau)\mathrm{d}\tau \right) e^{i2\pi f t}\mathrm{d}t \tag{2.5}$$

where, γ is the gyromagnetic ratio. $F(f)$ can be intuitively understood as the Fourier transform of the gradient integral. It can be seen from formulas 2.4 and 2.5 that signal attenuation in the OGSE sequence is only sensitive to the frequency component present in $F(f)$.

For such a gradient waveform, the diffusion time can generally be defined as:

$$\Delta_{\text{eff}} \approx 1/(4f) \tag{2.6}$$

where f is the reciprocal of the oscillation period (t_d). Measuring dMRI signals at different diffusion times enables the detection of time-dependent diffusion parameters and is used to characterize important microstructure properties such as cell size, intracellular fraction, and surface-volume ratio. See Sect. 1.3.

2.1.2 *Signal Readout Module*

(1) FSE

The FSE [3] sequence, also known as the Turbo Spin Echo (TSE) sequence, is a widely used signal acquisition method in dMRI, as shown in Fig. 2.4. Compared to the conventional spin echo sequence, the advantage of the FSE sequence lies in its ability to use multiple 180° RF pulses to acquire multiple echoes within a single TR period. By altering the phase-encoding gradient for each echo, multiple rows of *k*-space data can be collected. This allows for multiple signal acquisitions within one TR cycle, effectively reducing imaging time and enhancing acquisition efficiency.

Compared to the EPI sequence, the FSE sequence is less sensitive to changes in the B0 magnetic field. The multiple 180° RF pulses used in the FSE sequence can compensate for dephasing caused by diffusion gradients or magnetic field inhomogeneities, allowing the signal to refocus at the TE moment. Moreover, these refocusing pulses slow down signal decay (compared to the T2* decay rate in EPI), which is beneficial for achieving high-resolution and large field-of-view (FOV) imaging. Studies have shown [7] that the diffusion-weighted signal acquisition in FSE sequences can minimize the impact of magnetic susceptibility and eddy current artifacts on the images. The geometric accuracy of the acquired images is consistent with that of T2-weighted images, which is advantageous for the precise localization and diagnosis of lesions. The high signal retention capability of the FSE sequence also makes it a preferred option for high-field-strength scanning, enabling diffusion-weighted imaging of the lungs [8].

However, the FSE sequence also has some drawbacks. First, compared to the EPI sequence, the FSE sequence has a longer echo spacing, which results in longer acquisition times for the same number of echoes. This makes it more susceptible to motion artifacts. Second, the use of multiple 180° RF pulses in the FSE sequence can reduce contrast in certain regions of the image, leading to non-uniform signal intensity distribution. Finally, the FSE sequence typically requires a longer TE, which enhances T2 decay effects, especially in tissues with high water content. A longer TE

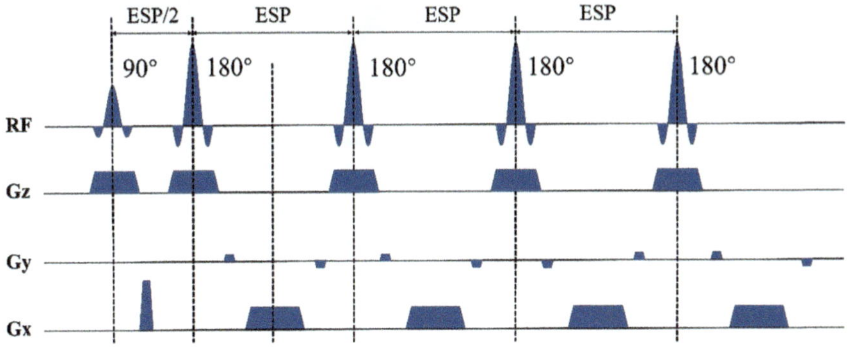

Fig. 2.4 Schematic diagram of FSE sequence frame

also increases the likelihood of susceptibility artifacts at tissue interfaces. Despite these challenges, the FSE sequence remains an extremely important and effective technique in dMRI imaging.

(2) EPI

With the continuous increase in magnetic field gradient strength and advancements in image acquisition technology, modern MRI machines are now capable of acquiring a single MRI slice within a very short time frame (50–100 ms). This significantly reduces the impact of patient motion on image quality. This imaging mode is commonly referred to as the EPI [4] sequence, as shown in Fig. 2.5. Following a single 90° radiofrequency (RF) pulse excitation, an oscillating frequency-encoding gradient and a low-amplitude phase-encoding gradient that switches intermittently are applied simultaneously. The gradient echo signals are read out during the oscillation of the frequency-encoding gradient to collect data. The initial frequency and phase-encoding gradients applied impart a negative phase shift to all spins, moving the starting point of signal acquisition from the center of *k*-space to the lower left corner. Subsequent positive and negative frequency-encoding gradients then scan *k*-space from left to right and right to left, respectively. Meanwhile, low-amplitude phase-encoding gradient pulses are continuously applied along the phase direction. The end result is that *k*-space is traversed in a "zigzag" pattern.

The EPI sequence is a commonly used technique in dMRI imaging, renowned for its ultra-fast imaging speed. However, the EPI sequence has some inherent issues regarding the quality of the images it acquires. First, the non-uniform distribution of the B_0 magnetic field in the scanned area can cause geometric distortions in the images acquired by the EPI sequence, especially at the edges of the scanned region. This can affect the localization and diagnosis of lesions. Second, the rapid switching of gradient pulses during the EPI sequence scan can induce eddy currents, which can cause geometric distortions and non-uniform signal intensity in the images, thereby reducing image quality. Lastly, because the EPI sequence collects data for the entire

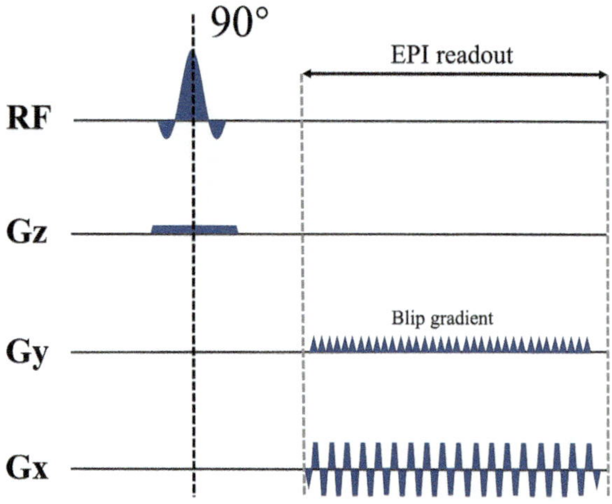

Fig. 2.5 Schematic diagram of EPI sequence framework

image in each echo acquisition, it can lead to signal dephasing within image voxels, reducing the actual image resolution. Despite these challenges, the EPI sequence, thanks to its fast imaging capabilities, still holds significant advantages in clinical applications and is highly favored by both doctors and researchers.

(3) GRASE

The GRASE [5] sequence is a specialized and highly efficient imaging technique in MRI that cleverly integrates the advantages of both Fast Spin Echo (FSE) and Echo Planar Imaging (EPI) sequences. In the GRASE sequence, a series of RF (radiofrequency) pulses generates and records a sequence of alternating gradient echoes and spin echoes. The process begins with a 90° RF pulse that excites the tissue signal, followed by several 180° refocusing pulses that produce multiple spin echoes. Around each spin echo, rapid switching of the readout gradient polarity generates multiple gradient echoes. The acquired signals contain information from both spin echoes and gradient echoes, enhancing sampling efficiency and image stability. A typical GRASE sequence is shown in Fig. 2.6.

The core strengths of the GRASE sequence lie in its integration of the benefits of FSE and EPI sequences. Compared to FSE sequences, GRASE can acquire more echoes in the same amount of time, improving sampling efficiency. Additionally, it requires fewer RF refocusing pulses, resulting in a lower specific absorption rate (SAR), which is particularly important for long-duration scans at high magnetic field strengths. Compared to EPI sequences, GRASE significantly reduces the phase accumulation of non-resonant spins using multiple RF refocusing pulses, effectively minimizing image distortion and artifacts, and enhancing image quality. These advantages make the GRASE sequence highly suitable for high-resolution and high-contrast imaging, especially in its ability to enhance the details of brain imaging.

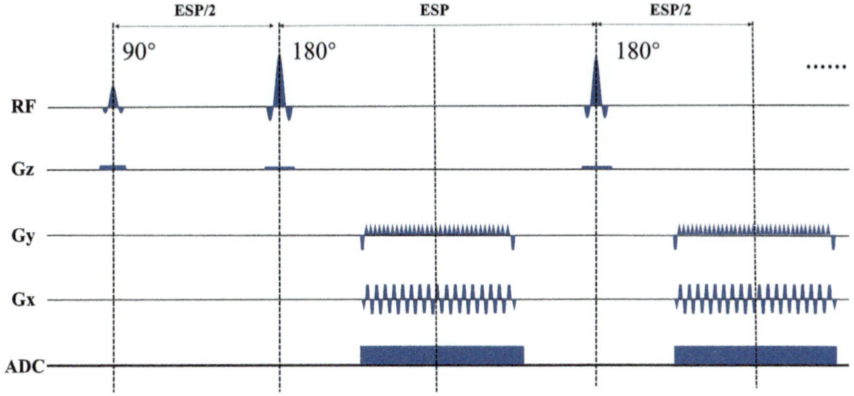

Fig. 2.6 Schematic diagram of GRASE sequence framework

2.2 High Resolution Diffusion MRI Sequence

2.2.1 3D Diffusion Imaging

Although 2D EPI (echo planar imaging) acquisition techniques have significantly improved the efficiency of data collection, they still face challenges such as long scan times and low signal-to-noise ratio (SNR) when acquiring high isotropic resolution whole-brain dMRI data. Finding solutions to these issues has become increasingly important in neuroscience research. In recent years, the development of various 3D dMRI acquisition schemes [9–11] has dramatically changed this situation by not only increasing the speed of dMRI data acquisition but also improving image SNR. In theory, the SNR advantage of 3D versus 2D is sqrt(N), where N is the number of slice direction encoding steps.

Another major advantage of 3D imaging technology is that it defines thinner slices in two directions using gradient encoding (i.e., 3D k-space), with slice-to-slice clarity similar to that within the plane. In contrast, 2D imaging often suffers from distorted layers and slice "cross-talk" between adjacent slices in 2D imaging acquired with a short TR, resulting in poorer inter-slice clarity and SNR loss. Additionally, the effective inter-slice resolution in 2D imaging primarily depends on the fidelity of the slice profile of the RF pulse, which is not ideal in actual imaging processes, leading to a decrease in inter-slice resolution.

Of course, 3D dMRI imaging technology also presents significant challenges. Since the 3D *k*-space encoding process in whole-brain imaging can only be completed through multi-segment acquisition, it is necessary to correct phase errors caused by motion before merging the segments. Generally speaking, phase errors exhibit spatial variations in all directions. Therefore, to comprehensively correct phase errors, navigator guidance is required in 3D space.

2.2.2 Multi-slab Acquisition

A primary goal in diffusion MRI has been the acquisition of high-resolution, isotropic images across the entire brain. However, the conventional 2D single-shot echo-planar imaging (ss-EPI) method faces a fundamental bottleneck: poor signal-to-noise ratio (SNR) efficiency. To achieve full brain coverage with thin, isotropic slices, 2D ss-EPI sequences require a long repetition time (TR) to accommodate all the slices, creating significant "dead time" where no signal is acquired. This leads to a low SNR per unit of scan time, limiting the achievable image quality. The diffusion-weighted 3D multi-slab [9] technique was developed specifically to overcome this limitation, offering a paradigm with markedly superior SNR efficiency for demanding diffusion studies.

The multi-slab approach achieves its efficiency by fundamentally changing the volume coverage strategy. Instead of acquiring many individual thin slices, the technique partitions the brain volume into a smaller number of thick, 3D-encoded "slabs". This design decouples the TR from the number of slices required for isotropic resolution, allowing the sequence to operate with an SNR-optimal TR of approximately 1-2 seconds. The TR of 3D multi-slab acquisition is proportional to the number of slabs, which can be adjusted to achieve optimal SNR efficiency [12] as fig. 2.7 depicted. For a given total scan time, this efficiency gain translates directly into a substantial increase in SNR, which enables the clear visualization of fine anatomical structures and provides higher quality data for quantitative analysis and tractography. Furthermore, the use of 3D Fourier encoding provides truly isotropic voxels with sharp, well-defined boundaries, avoiding the partial volume effects and degraded slice profiles inherent to 2D RF-pulse-defined slices.

Multi-shot 3D acquisition is inherently sensitive to motion-induced phase errors, which are particularly severe in dMRI and can cause significant artifacts if uncorrected. The multi-slab design provides an elegant solution to this challenge as well. By limiting the acquisition to relatively thin slabs, the complex phase variations along the through-slab (z) direction are constrained. This allows the 3D phase corruption to be accurately approximated and corrected using a simpler 2D navigator echo, a method that has been proven robust in numerous studies. While challenges such as artifacts at the slab boundaries exist, they can be effectively managed with techniques like slab overlap, optimized RF pulses, and post-processing corrections [13]. In summary, the 3D multi-slab technique provides a robust and efficient pathway to obtaining high-quality, high-SNR, and truly isotropic diffusion data across the whole brain.

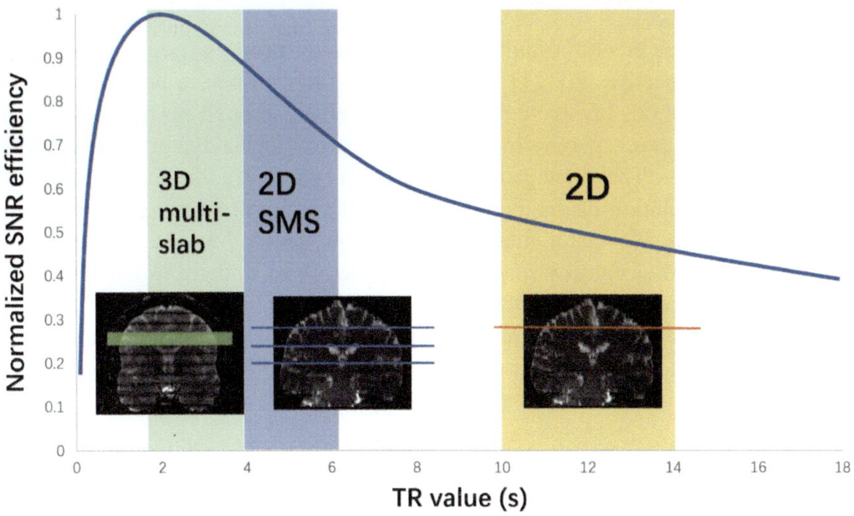

Fig. 2.7 SNR efficiency between 3D and 2D acqusition

2.2.3 Multi-segment Acquisition

In the field of dMRI, the strategy of multi-segment acquisition [14] offers significant advantages. It reduces the acquisition time required after each RF pulse excitation, thereby enhancing the signal-to-noise ratio (SNR) of the images and effectively mitigating the blurring effects caused by spin dephasing and T2* decay. This is particularly important because long acquisition times are often a necessary trade-off for high-resolution imaging [15], and as image resolution increases, the benefits of multi-segment acquisition become more pronounced.

Depending on how *k*-space is segmented, multi-segment acquisition can be divided into methods such as readout-segment and interleaved-segment acquisitions. Interleaved-segment acquisition is a common strategy. Its core idea is to divide the entire *k*-space into multiple segments along the phase-encoding direction. These segments are then acquired in an interleaved manner, with only one segment being collected after each RF pulse excitation, rather than acquiring the entire *k*-space at once. This segmented acquisition strategy distributes the acquisition burden evenly across the entire scanning process, as each excitation only requires the collection of a portion of the *k*-space data.

However, images acquired using multi-segment techniques are often susceptible to phase errors, especially in the presence of strong diffusion gradients. In contrast, single-shot acquisitions are immune to these phase errors because the entire image is acquired in a single excitation, with the phase information discarded and the diffusion effect reflected in the signal amplitude. Multi-segment acquisition, on the other hand, must retain phase information to accurately combine the different *k*-space segments obtained from each excitation. Failure to correct for motion-induced phase changes before merging these *k*-space segments can result in aliasing artifacts in the image, as shown in Fig. 2.8.

To correct phase errors in multi-segment diffusion imaging, both retrospective [16–19] and prospective [20, 21] phase correction methods have been developed. Retrospective methods involve correcting phase errors through post-processing of the images, whereas prospective methods involve acquiring navigator signals during the imaging process. These navigator signals typically require additional *k*-space data to create low-resolution navigator images, which are used to correct phase errors between segments. Navigator images provide undistorted, low-resolution images corresponding to the central *k*-space region of the image, and are usually acquired immediately before or after the imaging data, or extracted directly from the imaging data itself (self-navigation technique). A typical pulse sequence incorporating navigators is shown in Fig. 2.9. In cases with a large number of segments, prospective phase error correction methods are generally more effective.

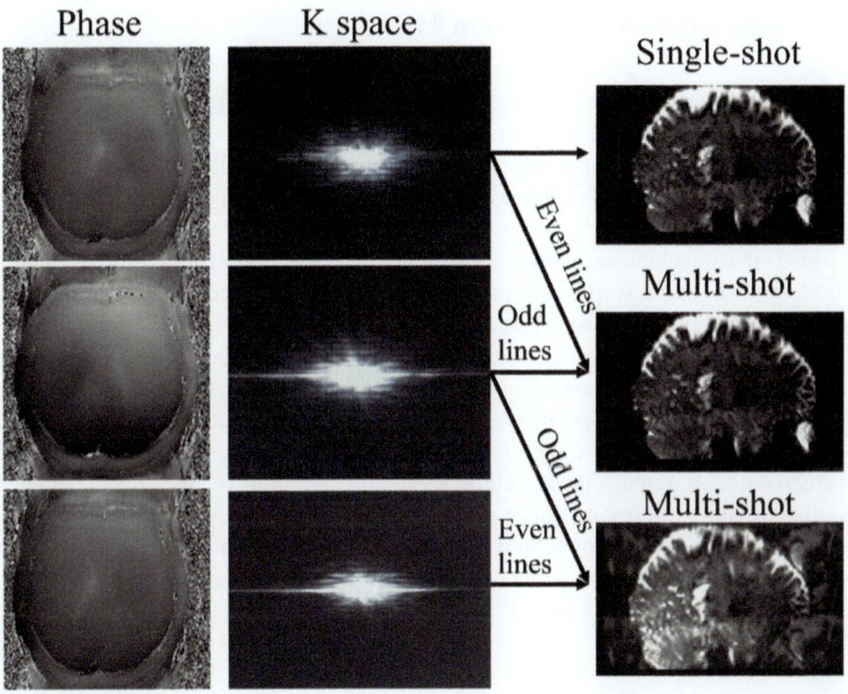

Fig. 2.8 Aliasing artifacts. During the diffusion encoding process, subject motion can cause severe spatial phase shifts. However, this does not affect single-shot acquisitions (**a**). In contrast, for multi-segment acquisitions, the impact of phase inconsistencies is particularly pronounced. For example, if the subject remains stable during the first two TR periods (no phase changes) but experiences head rotation during the third TR period (introducing a linear phase offset), the combination of even k-space lines from the first TR and odd k-space lines from the second TR will produce an artifact-free image (**b**). Conversely, combining odd k-space lines from the second TR with even k-space lines from the third TR will result in an image with severe artifacts (**c**)

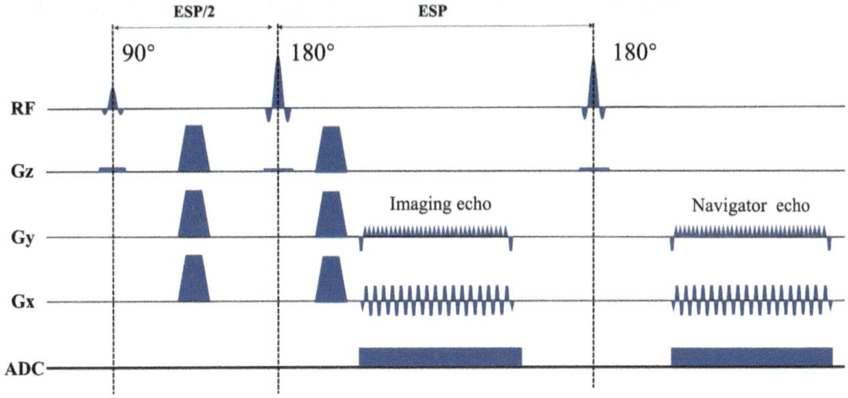

Fig. 2.9 Schematic diagram of navigator sequence frame

2.3 Acceleration of Diffusion MRI

Rapid imaging techniques represent a significant breakthrough in the field of dMRI, offering new possibilities for both basic scientific research and clinical diagnostics. These techniques include partial Fourier imaging [3] and parallel imaging [3], among others. The advantages of rapid dMRI techniques are primarily reflected in three key areas.

First, these techniques reduce the number of k-space samples required, thereby shortening scan times. This not only minimizes patient discomfort during clinical scans but also enhances the efficiency of clinical workflows. Second, rapid imaging techniques maintain high image quality. Despite the acceleration, the use of sophisticated algorithms and advanced hardware ensures that high-resolution dMRI images are obtained without compromising image quality. This is particularly important for accurately depicting subtle anatomical structures and the microscopic movement of water molecules. Finally, these techniques significantly mitigate the impact of patient motion, thereby improving image fidelity.

Overall, rapid imaging techniques have transformed dMRI by making it faster, more efficient, and more reliable, while still delivering high-quality images that are essential for both research and clinical applications.

2.3.1 Partial Fourier Imaging

Partial Fourier imaging is a technique that has had a profound impact on dMRI scans, both in theoretical research and practical applications. It leverages the conjugate symmetry of k-space data, meaning that half of the k-space data (or slightly more) can theoretically reflect the entire k-space information. This symmetry allows for the collection of only half of the k-space data, with the remainder being inferred through symmetry principles. This approach significantly reduces the amount of data acquisition, thereby shortening overall scan time. Additionally, partial Fourier imaging offers flexibility in adjusting the sampling ratio according to specific needs and applications, providing high operational adaptability. As a result, this technique not only dramatically reduces scan time but also enhances image quality by lowering the echo time (TE) and increasing the signal-to-noise ratio (SNR). This meets the urgent needs of researchers and clinicians for rapidly obtaining high-quality dMRI images.

However, partial Fourier imaging also faces several challenges, primarily due to various errors encountered during actual scanning processes. These include inhomogeneities in the B_0 magnetic field, eddy currents, magnetic susceptibility effects, and physiological motion. All these factors can lead to the loss of conjugate symmetry in k-space, ultimately affecting the effectiveness of partial Fourier imaging. Therefore, it is necessary to address these issues through specific algorithms for processing and correction to ensure optimal scan results from partial Fourier imaging. Despite these challenges, partial Fourier imaging remains an important technique frequently used in clinical settings and has demonstrated significant potential in various dMRI applications. With advancements in algorithms and MRI hardware technology, this technique is poised to play an increasingly important role, bringing greater convenience to both basic scientific research and clinical healthcare.

2.3.2 *Parallel Imaging*

The application of parallel imaging techniques [22] in dMRI represents a significant milestone in the evolution of MRI technology. By simultaneously acquiring dMRI signals with multiple receiver coils, this approach has achieved remarkable improvements in both imaging speed and image quality. Techniques such as Sensitivity Encoding (SENSE) [23] and GRAPPA [24] are prominent examples within this category. The fundamental principle involves multiple coils, distributed in different spatial locations, acquiring data concurrently and then reconstructing the image using advanced algorithms. Each coil covers a specific region of the imaging area and has a unique sensitivity profile. These profiles provide additional spatial information, enabling the reconstruction of a complete image from fewer *k*-space samples. Parallel imaging not only accelerates the imaging process and significantly reduces dMRI scan times but also enhances the signal-to-noise ratio (SNR) and spatial resolution of the images. This improvement allows for better visualization of subtle tissue microstructures, such as the architecture of nerve fibers.

Moreover, parallel imaging techniques mitigate image artifacts caused by involuntary patient motion, which is particularly beneficial for pediatric patients or individuals who have difficulty remaining still for extended periods. In terms of application, parallel imaging has propelled research across various fields, including neuroscience and oncology.

(1) SENSE

SENSE is a crucial strategy in parallel imaging techniques. It leverages the spatial variations in coil sensitivity to reconstruct images. When using SENSE, the process begins with a prescan to obtain sensitivity maps for each receiver coil. During the actual imaging, multiple receiver coils simultaneously collect signals, each acquiring a portion of the *k*-space data. By exploiting the spatial differences in coil sensitivity, each coil provides unique information about the entire image, even though only a fraction of the *k*-space data is sampled. This information is then combined with the pre-acquired coil sensitivity maps to reconstruct a complete, high-quality image using the SENSE algorithm.

SENSE significantly reduces the number of *k*-space samples required, thereby enhancing imaging speed and spatial resolution. However, a major challenge with SENSE is the amplification of image noise, which can degrade image quality. Additionally, the quality of images obtained using SENSE is highly dependent on the distribution and performance of the receiver coils. In other words, the effectiveness of SENSE relies heavily on the accurate acquisition of coil sensitivity information. This places stringent demands on the measurement techniques and calibration algorithms for coil sensitivity and requires researchers to optimize the algorithms and parameters.

(2) GRAPPA

GRAPPA is a parallel imaging technique that directly interpolates missing data in *k*-space and has been widely applied in dMRI. Specifically, GRAPPA uses multiple receiver coils to simultaneously acquire partial *k*-space data and then predicts and fills in the missing *k*-space data based on the spatial relationships between the coils. In this process, GRAPPA first extracts the unique patterns of each coil from the acquired data and uses these patterns to estimate the missing *k*-space data. This method significantly reduces scan time while maintaining image quality. Figure 2.10 illustrates the specific steps for a four-fold accelerated scan using GRAPPA. First, GRAPPA coefficients are obtained through autocalibration signals (ACS), and then these coefficients are applied to the undersampled data to fill in the missing *k*-space data.

In the field of dMRI, GRAPPA has been highly favored by researchers due to its multiple advantages. First, GRAPPA has strong acceleration capabilities and can achieve high-fold accelerated imaging, especially in systems equipped with multi-channel coils. Second, through a rational reconstruction strategy, GRAPPA can maintain high image quality while increasing imaging speed. Finally, compared to SENSE, GRAPPA is less sensitive to magnetic susceptibility-induced field distortions.

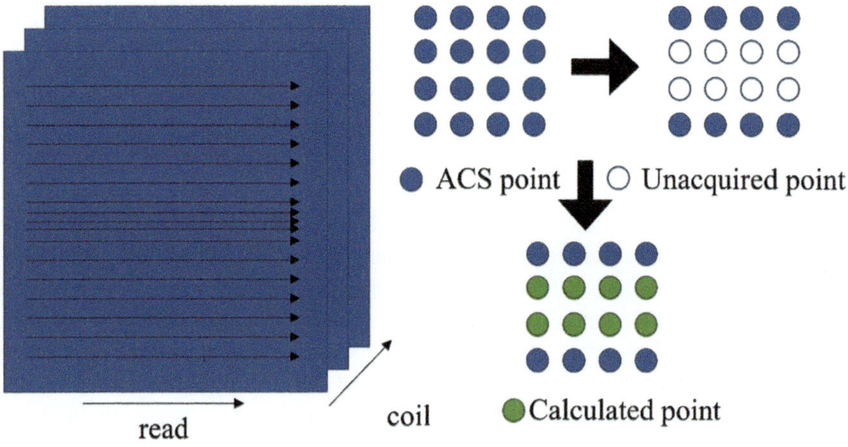

Fig. 2.10 Schematic diagram of reconstruction by GRAPPA. The acceleration factor is set to 4

2.4 Conclusion

This chapter provides a comprehensive introduction to the core theories and sequence techniques of dMRI imaging. It begins with an in-depth exploration of dMRI sequence design, covering various diffusion preparation methods and data acquisition approaches. Building on this foundation, the chapter then delves into high-resolution and high signal-to-noise ratio (SNR) dMRI techniques. Finally, it details the principles and implementation methods of fast dMRI imaging techniques.

References

1. Neeman M, Freyer JP, Sillerud LO (1990) Pulsed-gradient spin-echo diffusion studies in NMR imaging. Effects of the imaging gradients on the determination of diffusion coefficients. J Magnet Reson 90(2):303–312
2. Baron CA, Beaulieu C (2014) Oscillating gradient spin-echo (OGSE) diffusion tensor imaging of the human brain. Magnet Reson Med 72(3):726–736
3. Sarlls JE, Pierpaoli C (2008) Diffusion-weighted radial fast spin-echo for high-resolution diffusion tensor imaging at 3T. Magnet Reson Med Off J Int Soc Magnet Reson Med 60(2):270–276
4. Setsompop K, Cohen-Adad J, Gagoski BA et al (2012) Improving diffusion MRI using simultaneous multi-slice echo planar imaging. Neuroimage 63(1):569–580
5. Wu D, Liu D, Hsu YC et al (2021) Diffusion-prepared 3D gradient spin-echo sequence for improved oscillating gradient diffusion MRI. Magnet Reson Med 85(1):78–88
6. Stejskal EO, Tanner JE (1965) Spin diffusion measurements: spin echoes in the presence of a time-dependent field gradient. J Chem Phys 42(1):288–292
7. Tyagi N, Cloutier M, Zakian K et al (2019) Diffusion-weighted MRI of the lung at 3T evaluated using echo-planar-based and single-shot turbo spin-echo-based acquisition techniques for radiotherapy applications. J Appl Clin Med Phys 20(1):284–292
8. Sigmund EE, Gutman DJMRI (2011) Diffusion-weighted imaging of the brain at 7 T with echo-planar and turbo spin echo sequences: preliminary results. Magnet Reson Imag 29(6):752–765
9. Engström M, Skare S (2013) Diffusion-weighted 3D multislab echo planar imaging for high signal-to-noise ratio efficiency and isotropic image resolution. Magnet Reson Med 70(6):1507–1514
10. Frank LR, Jung Y, Inati S et al (2010) High efficiency, low distortion 3D diffusion tensor imaging with variable density spiral fast spin echoes (3D DW VDS RARE). Neuroimage 49(2):1510–1523
11. Jeong EK, Kim SE, Kholmovski EG et al (2006) High-resolution DTI of a localized volume using 3D s ingle-s hot diffusion-weighted ST imulated e cho-p lanar i maging (3D ss-DWSTEPI). Magnet Reson Med Off J Int Soc Magnet Reson Med 56(6):1173–1181
12. Wu W, Poser BA, Douaud G et al (2016) High-resolution diffusion MRI at 7T using a three-dimensional multi-slab acquisition. NeuroImage 143:1–14
13. Wu W, Koopmans PJ, Frost R et al (2016) Reducing slab boundary artifacts in three-dimensional multislab diffusion MRI using nonlinear inversion for slab profile encoding (NPEN). Magnet Reson Med 76(4):1183–1195
14. Madore B, Chiou JYG, Chu R et al (2014) Accelerated multi-shot diffusion imaging. Magnet Reson Med 72(2):324–336
15. Johansen-Berg H, Behrens TE (2013) Diffusion MRI: from quantitative measurement to in vivo neuroanatomy. Academic Press, New York

16. Chen N-K, Guidon A, Chang H-C et al (2013) A robust multi-shot scan strategy for high-resolution diffusion weighted MRI enabled by multiplexed sensitivity-encoding (MUSE). Neuroimage 72:41–47
17. Mani M, Jacob M, Kelley D et al (2017) Multi-shot sensitivity-encoded diffusion data recovery using structured low-rank matrix completion (MUSSELS). Magnet Reson Med 78(2):494–507
18. Lobos RA, Kim TH, Hoge WS et al (2018) Navigator-free EPI ghost correction with structured low-rank matrix models: new theory and methods. IEEE Trans Med Imag 37(11):2390–2402
19. Chang HC, Hui ES, Chiu PW et al (2018) Phase correction for three-dimensional (3D) diffusion-weighted interleaved EPI using 3D multiplexed sensitivity encoding and reconstruction (3D-MUSER). Magnet Reson Med 79(5):2702–2712
20. Gumus K, Keating B, Poser BA et al (2014) Prevention of motion-induced signal loss in diffusion-weighted echo-planar imaging by dynamic restoration of gradient moments. Magnet Reson Med 71(6):2006–2013
21. Herbst M, Maclaren J, Weigel M et al (2012) Prospective motion correction with continuous gradient updates in diffusion weighted imaging. Magnet Reson Med 67(2):326–338
22. Deshmane A, Gulani V, Griswold MA et al (2012) Parallel MR imaging. J Magnet Reson Imag 36(1):55–72
23. Pruessmann KP, Weiger M, Scheidegger MB et al (1999) SENSE: sensitivity encoding for fast MRI. Magnet Reson Med Off J Int Soc Magnet Reson Med 42(5):952–962
24. Griswold MA, Jakob PM, Heidemann RM et al (2002) Generalized autocalibrating partially parallel acquisitions (GRAPPA). Magnet Reson Med Off J Int Soc Magnet Reson Med 47(6):1202–1210

Chapter 3
3D Diffusion Gradient Prepared Gradient and Spin Echo Imaging

Abstract This chapter introduces a novel 3D oscillating gradient prepared gradient and spin echo (3D OGprep-GRASE) sequence for short diffusion time time-dependent diffusion MRI (TDDMRI) on a clinical 3 T system. This sequence was developed to address limitations of conventional 2D oscillating gradient echo-planar imaging (OG-EPI), such as long scan times and low signal-to-noise ratio (SNR). Comparisons with 2D OG-EPI demonstrated that the 3D OGprep-GRASE sequence significantly reduced scan times and improved SNR, particularly for larger imaging volumes. The improved SNR led to better diffusion tensor imaging (DTI) reconstruction with reduced noise and image distortion. Additionally, a 3D pulsed gradient prepared gradient and spin echo (3D PGprep-GRASE) sequence was developed for long diffusion times, and both sequences were used to investigate the diffusion time dependence of diffusion parameters in human brain tissue, showing similar sensitivity to ADC changes as 2D EPI. The findings suggest that the 3D OGprep-GRASE sequence offers significant advantages for clinical TDDMRI applications, paving the way for improved brain disease diagnosis and tracking.

Keywords 3D OGprep-GRASE · TDDMRI · Diffusion MRI · Oscillating gradient · Pulsed gradient · GRASE · EPI · SNR

3.1 Introduction

As elucidated in Sect. 1.3 in Chap. 1, the diffusion effect of restricted water molecules varies across different microenvironments in biological tissues. dMRI technology can obtain crucial information about tissue microstructure by detecting these restricted water molecule diffusion properties. Research indicates that the signal intensity of dMRI depends not only on the gradient direction and diffusion weighting *b*-value [1–5] but also on the diffusion time [6–8] and relaxation time [9, 10]. More specifically, diffusion time determines the sensitivity of dMRI signals to microstructural features at different spatial scales [11, 12]. Building upon this, the diffusion time-dependent characteristics of different tissues can be obtained by measuring dMRI signals at

D. Wu et al., *Development and Application of 3D Diffusion Magnetic Resonance Imaging Sequences*, Advanced Topics in Science and Technology in China 5,
https://doi.org/10.1007/978-981-95-0808-2_3

various diffusion times, known as TDDMRI signals. This characteristic information, including cell size and cell composition ratios, is essential for understanding and delineating the microstructural properties of tissues.

As discussed in Sect. 1.4 in Chap. 1, the emergence [8] of oscillating gradient diffusion magnetic resonance imaging (OG-dMRI) acquisition technology was a landmark breakthrough because conventional pulsed gradient diffusion magnetic resonance imaging (PG-dMRI) sequences could not achieve very short diffusion times. OG encoding gradients typically consist of multiple cosine gradient waveforms and have been widely used in preclinical systems for TDDMRI studies on animals [13–16]. However, this technique faces significant challenges in clinical applications due to its high requirements for gradient strength and gradient switching rates. On the one hand, because a single oscillation cycle provides a limited *b*-value, multiple oscillation cycles are often needed to reach a usable *b*-value. This inevitably lengthens the duration of the diffusion encoding gradient, leading to a longer echo time (TE) and a decrease in image signal-to-noise ratio (SNR). On the other hand, the use of strong oscillating gradients can cause high duty cycles and overheating of the MRI scanner, necessitating an increase in repetition time (TR) to reduce the duty cycle and dissipate heat through the cooling system. In 2D multi-slice dMRI sequences, the need to repeatedly apply strong gradients for each slice, coupled with the linear increase in TR with the number of slices, results in long scan times for OG-dMRI with high resolution or a large field of view (FOV). Although these issues specifically pertain to OG-dMRI, the need to maintain the same TE and TR in both PG and OG scans for subsequent diffusion time dependency analysis means that these problems urgently need to be addressed for TDDMRI research.

Many studies have developed various 3D dMRI techniques to improve image SNR and resolution [17–20]. Among these, 3D gradient and spin echo (GRASE) acquisition combines the advantages of 3D fast spin echo (FSE) and 3D echo-planar imaging (EPI) acquisitions, demonstrating efficient 3D encoding acquisition efficiency [21] and insensitivity to field strength variations [22] in dMRI. Therefore, this study designed a novel 3D oscillating gradient prepared gradient and spin echo (3D OGprep-GRASE) sequence based on 3D GRASE acquisition for short diffusion time TDDMRI acquisition to address the aforementioned issues in OG-dMRI. This sequence was compared with the conventional 2D OG-EPI sequence on a clinical 3 T system in terms of scan time, SNR, image distortion, and tensor reconstruction quality. This study also designed a 3D pulsed gradient prepared gradient and spin echo (3D PGprep-GRASE) sequence for long diffusion time TDDMRI acquisition. Both sequences were used to investigate the diffusion time dependence of diffusion parameters in human brain gray and white matter.

3.2 Theory

As shown in Figs. 3.1 and 3.2, the 3D OGprep-GRASE and 3D PGprep-GRASE imaging sequences designed in this study are mainly composed of four modules: a global stabilization module, a diffusion preparation module, a fat saturation module, and a GRASE readout module. The difference between these two sequences lies in the different gradients added in the diffusion preparation module. The following will introduce each module separately.

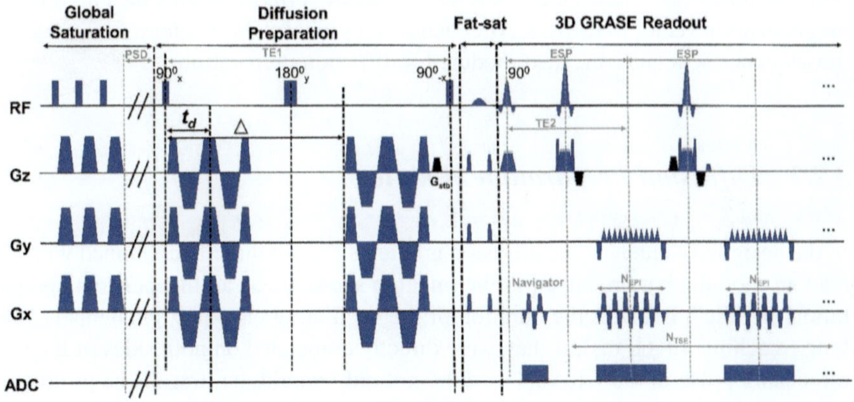

Fig. 3.1 Schematic diagram of the 3D OGprep-GRASE sequence framework. This sequence includes a global stabilization module, an oscillating gradient diffusion preparation module, a fat saturation module, and a GRASE acquisition module

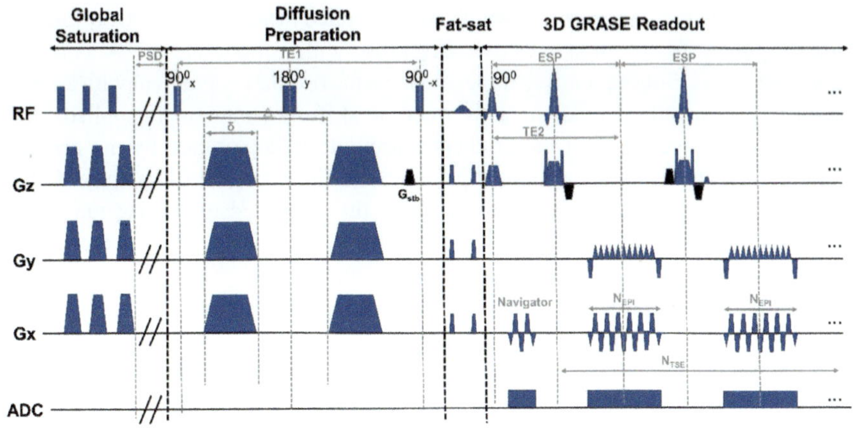

Fig. 3.2 Schematic diagram of the 3D PGprep-GRASE sequence framework. This sequence includes a global stabilization module, a pulsed gradient diffusion preparation module, a fat saturation module, and a GRASE acquisition module

3.2.1 Global Stabilization Module

The global stabilization module is added at the beginning of the sequence, and its purpose is to remove the residual magnetization in the transverse plane from the previous TR. At the start of the sequence scan, the global stabilization module first applies a B1 radiofrequency pulse, and then simultaneously applies gradients with completely equal amplitudes in the x, y, and z directions. By repeating the same process three times, it is ensured that the residual transverse magnetization signal after the previous TR signal acquisition is completely eliminated. Next, the system will wait for a period called the post saturation delay (PSD), allowing the longitudinal magnetization vector to recover. After achieving global stability and sufficient delay, the sequence will enter the next module for diffusion preparation.

3.2.2 Diffusion Preparation Module

In the designed sequence, the diffusion gradients are not directly combined with the readout module, but are added to the imaging sequence as an independent "preparation module". Because the duration of the oscillating encoding gradients is very long (reaching 40–60 ms), if they were directly embedded on both sides of the first refocusing pulse in the GRASE module, not only would the first TE be prolonged to about 150 ms, but also, to satisfy the Carr-Purcell-Meiboom-Gill (CPMG) conditions, the echo spacing between subsequent refocusing pulses must be equal, which would lead to rapid signal decay. Therefore, this strategy of separating diffusion encoding from image acquisition is crucial for OG-dMRI imaging.

In the diffusion preparation module, a pair of trapezoidal cosine oscillating gradients or pulsed gradients are used, which are embedded in $90°_x$–$180°_y$–$90°_{-x}$ radiofrequency pulses. First, a non-selective hard pulse with a flip angle of 90° is applied along the x-axis. Subsequently, a trapezoidal cosine oscillating gradient with a certain frequency or a pulsed gradient with a certain diffusion interval is applied in the set diffusion direction, and the gradient amplitude is determined by the preset b-value. Then, an adiabatic orthogonal pulse with a flip angle of 180° is applied along the y-axis, the purpose of which is to allow the dispersed transverse magnetization components to refocus. Next, the same diffusion gradients as before are applied, with completely identical amplitude, waveform, and duration. Subsequently, a spoiler gradient (G_{stb}) is applied along the z-axis and set to the same amplitude as the crusher gradients in the subsequent GRASE acquisition module, the purpose of which is to convert amplitude errors caused by motion or eddy currents into phase errors, thereby reducing signal modulation related to phase errors [23]. Finally, a non-selective hard pulse with a flip angle of 90° is applied along the $-x$ axis to convert the transverse magnetization vector into the longitudinal magnetization vector. Among these, the echo time in the diffusion preparation module is denoted as TE_1.

3.2.3 Fat Saturation Module

A fat saturation module is also added to the imaging sequence. In this stage, a frequency-selective Gaussian pulse with a flip angle of 110° is applied to convert most of the longitudinal magnetization vector of fat into the transverse magnetization vector. Then, gradients are simultaneously applied in the x, y, and z directions to completely eliminate the transverse magnetization vector of fat. The main role of the fat saturation module is to suppress fat signals in the image, thereby obtaining higher quality image data.

3.2.4 Gradient Spin Echo Acquisition Module

The 3D GRASE acquisition module is implemented through EPI encoding in the first phase-encoding direction and FSE encoding in the second phase-encoding direction. In the EPI direction, the number of phase encodings is represented by N_{EPI}, while for the FSE direction, the number of phase encodings is represented by N_{FSE}. It is worth noting that a center-out encoding strategy is adopted in the FSE direction. This means that after the first refocusing pulse, EPI acquisition is first used to fill the central position of the k-space in the k_z direction. This is a strategy to effectively improve image SNR. To improve image quality and stability, spoiler gradients and crusher gradients are simultaneously applied along the z-axis before and after each EPI readout, as well as on both sides of the refocusing pulses. In the GRASE acquisition module, the echo time is defined as TE_2, which is equal to the echo spacing for signal acquisition in FSE mode. Therefore, the total TE obtained is the sum of TE_1 and TE_2. By segmenting the phase encoding in the EPI direction for readout, the acquisition efficiency and image quality can be improved.

3.3 Experimental Methods

3.3.1 Magnetic Resonance Experiments

This study conducted a total of two sets of experiments. Experiment 1 set two comparison objectives: first, to compare the scan times of the 3D OGprep-GRASE and 2D OG-EPI sequences; second, to compare the signal-to-noise ratio (SNR) of images scanned by these two sequences to determine the differences in their imaging performance. It is important to note that the SNR of 3D sequence scans is directly related to the imaging volume. Therefore, Experiment 1 set up OG-dMRI scanning experiments with different imaging ranges. The purpose of Experiment 2 was to investigate the time dependence of diffusion parameters in human brain gray and white matter.

Single-shot 3D OGprep-GRASE and 3D PGprep-GRASE sequences were used to collect DWI at short and long diffusion times, respectively.

All experiments in this study were performed on a 3 T MAGNETOM Prisma Siemens MRI scanner equipped with a 64-channel head coil. Seven healthy volunteers were recruited for this study (average age = 23 ± 2 years, range 20–25 years, 5 males, 2 females). Before participating in the experiments, all volunteers fully understood the purpose and procedure of the study and signed an informed consent form. All research procedures were approved by the ethics review committee of the affiliated institution.

3.3.2 Data Acquisition

Experiment 1 first used 3D OGprep-GRASE and PGprep-GRASE sequences with a *b*-value of 50 s/mm^2 to image a 2% agar gel phantom and calibrate the measured ADC. In human brain imaging, OG dMRI was performed with an oscillation frequency of 50 Hz, three oscillation cycles, $b = 500$ s/mm^2, 12 diffusion directions, and one non-diffusion-weighted (b0) imaging parameter. Both 2D and 3D sequences were set with a matrix size of 80 × 80, a FOV of 220 × 220 mm^2, an in-plane resolution of 2.75 × 2.75 mm^2, a slice thickness of 3 mm, a TE of 158 ms, and two acquisitions were averaged. The following three ranges of imaging schemes were adopted:

(1) 30 mm slice imaging range: Single-shot 3D OGprep-GRASE sequence acquisition, $N_{\mathrm{EPI}} = 59$ (partial Fourier factor = 74%), $N_{\mathrm{FSE}} = 12$ (20% slice oversampling), TR = 3000 ms, scan time was 1.24 min. 2D OG-EPI sequence acquisition, $N_{\mathrm{EPI}} = 60$ (partial Fourier factor = 75%), 10 slices, TR = 4200 ms (minimum TR), scan time was 1.53 min.

(2) 60 mm slice imaging range: Two-shot 3D OGprep-GRASE sequence acquisition, $N_{\mathrm{EPI}} = 29$, $N_{\mathrm{FSE}} = 24$, scan time was 2.42 min, other parameters were the same as in scheme (1). 2D OG-EPI sequence acquisition, 20 slices, TR = 8400 ms (minimum TR), scan time was 3.44 min, other parameters were the same as in scheme (1).

(3) 90 mm slice imaging range: Three-shot 3D OGprep-GRASE sequence acquisition, $N_{\mathrm{EPI}} = 21$, $N_{\mathrm{FSE}} = 36$, scan time was 4 min, other parameters were the same as in scheme (1). 2D OG-EPI sequence acquisition, 30 slices, TR = 12,600 ms (minimum TR), scan time was 5.53 min, other parameters were the same as in scheme (1).

This experiment also performed two-shot 3D PGprep-GRASE and OGprep-GRASE sequence acquisitions to test image quality, with $\delta = 20$ ms, $\Delta = 30$ ms, and all other imaging parameters the same as in scheme (2).

In Experiment 2, the single-shot 3D OGprep-GRASE sequence was acquired at oscillation frequencies of 25 (one cycle) and 50 Hz (two cycles), $b = 500$ s/mm^2, six directions were acquired for each, repeated twice, $\mathrm{TE_1/TE_2/TR} = 84/32/3000$ ms, matrix size = 80 × 80 × 10, resolution = 2.75 × 2.75 × 5 mm^3. The single-shot 3D

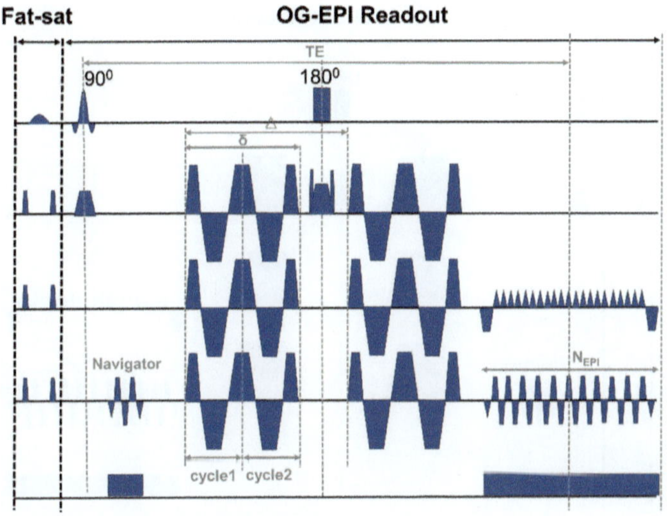

Fig. 3.3 Schematic diagram of the 2D OG-EPI sequence framework

PGprep-GRASE sequence was acquired with parameters $\delta = 20$ ms, $\Delta = 30$, and 60 ms, with other parameters matched to those in OGprep. As a comparison, this experiment acquired data using the 2D OG-EPI sequence at oscillation frequencies of 25 and 50 Hz, and using the 2D PG-EPI sequence with pulsed gradients at $\delta =$ 20 ms and $\Delta = 30/60$ ms, with other sequence parameters consistent with the 3D GRASE scan. Because fewer and less steeply ramped diffusion directions were used in Experiment 2, the same *b*-value as in Experiment 1 could be achieved at 50 Hz with fewer oscillation cycles.

The sequence diagrams for 2D OG-EPI and 2D PG-EPI are shown in Figs. 3.3 and 3.4, respectively.

3.3.3 Data Analysis

The multiplexed sensitivity encoding (MUSE) [24] method was used to correct phase errors in the 3D GRASE sequence data acquired with a small number of segments during large-volume imaging in Experiment 1. The MUSE method divides the originally large *k*-space dataset into a small number of segments using different phase encoding steps. Each segment samples *k*-space points in the first phase encoding direction using an interleaved acquisition approach. Because each segment has a smaller *k*-space, they are less susceptible to image distortions and experience less signal loss. After acquisition, the segments are reconstructed using SENSE technology, and then merged into a complete, less distorted image by solving an inverse reconstruction problem and an optimization problem.

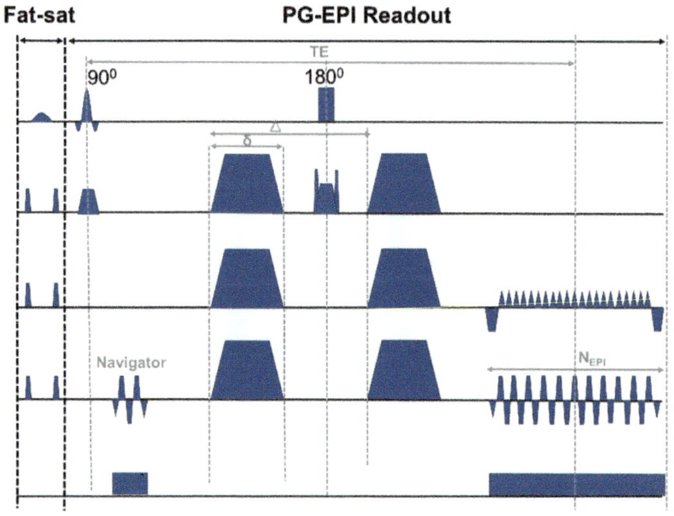

Fig. 3.4 Schematic diagram of the 2D PG-EPI sequence framework

MRtrix (https://www.mrtrix.org) software was used for image post-processing, including Gibbs artifact removal, eddy current correction, image registration, bias field correction, and diffusion tensor reconstruction. The image SNR was calculated by dividing the mean of two b_0 images by the standard deviation of their difference. Simultaneously, the sequence imaging quality was evaluated using DTI metrics such as ADC, FA, and direction-encoded color maps. Regional noise was assessed using the coefficient of variation (CV) [25], defined as the standard deviation of ADC or FA within a region divided by the mean. An experienced physician manually delineated the regions of interest (ROI) for this study: frontal and posterior frontal subcortical white matter, posterior limb of the internal capsule, basal ganglia, and thalamus.

Analysis of variance (ANOVA) and post-hoc *t*-tests were used to compare the SNR and CV of the 2D OG-EPI sequence and the 3D OGprep-GRASE sequence at different imaging volumes. Concurrently, ANOVA and post-hoc *t*-tests were also used to evaluate the diffusion time dependence effects of OG-dMRI and PG-dMRI sequences on different tissues (comparing ADC at different diffusion times) and the SNR differences between 3 and 2D sequences. All statistical analyses were performed using MATLAB (Mathworks, Inc., Natick, MA) software. For all results, a Bonferroni-corrected $p < 0.05$ was considered statistically significant.

3.4 Experimental Results

This study first calibrated the 3D GRASE sequence using an agar gel phantom. In the OG-dMRI images at 50 and 25 Hz, the measured average ADC values were 2.13 ± 0.04 and 2.10 ± 0.04 μm²/ms, respectively. In the PG-dMRI images at 30 and 60 ms, the obtained average ADC values were 2.11 ± 0.05 and 2.10 ± 0.04 μm²/ms, respectively. There was no statistically significant difference in ADC across these four different diffusion times (corresponding p-value was 0.28).

Figure 3.5 shows the comparison of images acquired by the 3D PGprep-GRASE and 3D OGprep-GRASE sequences before and after MUSE reconstruction. It can be seen that the PGprep-DWI images have obvious aliasing artifacts caused by inter-segment phase errors, and these errors can be effectively corrected by MUSE reconstruction. In contrast, the aliasing artifacts caused by inter-segment phase errors in OGprep-DWI images are much less.

As can be seen from the ADC, FA, and direction-encoded color maps in Fig. 3.6, the improved image SNR enabled the 3D OGprep-GRASE sequence to achieve better DTI reconstruction results. Although the noise in the DTI parameter maps of voxel-wise fitted signals in the 3D OGprep-GRASE data with a 30 mm slab imaging range was similar to that of the 2D OG-EPI data, the 3D OGprep-GRASE images gradually became clearer than the 2D OG-EPI images as the number of excitation segments increased or the slab imaging range expanded. For example, in the 60 and 90 mm slab imaging ranges, there was obvious noise and fitting errors in the direction-encoded

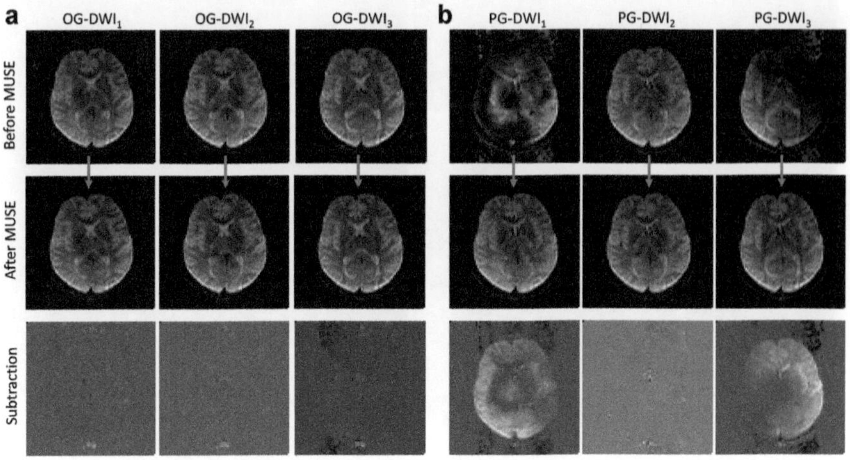

Fig. 3.5 MUSE Reconstruction Results. DWI images of three diffusion directions obtained by two-segment excitation OGprep (**a**) and PGprep (**b**) GRASE sequences, images before and after MUSE reconstruction, and difference images before and after reconstruction. The diffusion gradient directions of DWI1, DWI2, and DWI3 were, [−1, 0, 1], and, respectively. MUSE reconstruction can effectively correct phase errors in PG-DWI. For OG-DWI, there are almost no phase errors even without MUSE reconstruction [26]

color maps fitted by the 2D OG-EPI sequence data, but the noise and fitting errors were reduced in the color maps fitted by the 3D OGprep-GRASE sequence data (Fig. 3.6c). In addition, the image distortion in the orbitofrontal cortex in the 2D OG-EPI image (yellow arrow in Fig. 3.6d) was significantly improved in the 3D OGprep-GRASE image.

This study used 3D GRASE and 2D EPI sequences to obtain ADC measurements in several white matter and gray matter ROIs in OG-dMRI (50 and 25 Hz) and PG-dMRI (30 and 60 ms) data. The analysis showed that the ADC of both sequences exhibited significant ADC diffusion time dependence in white and gray matter regions ($p < 0.001$). ANOVA indicated no significant sequence differences, as shown in Fig. 3.7. This further confirmed the robustness of the sequence acquisition methods. In the white matter under the posterior frontal cortex, posterior limb of the internal capsule, and thalamus, this experiment observed a gradual decrease in ADC as the diffusion time increased from 50 to 25 Hz ($\Delta_{\mathrm{eff}} = 5$, 10 ms) to 30, 60 ms. More significant ADC differences between OG-dMRI and PG-dMRI were

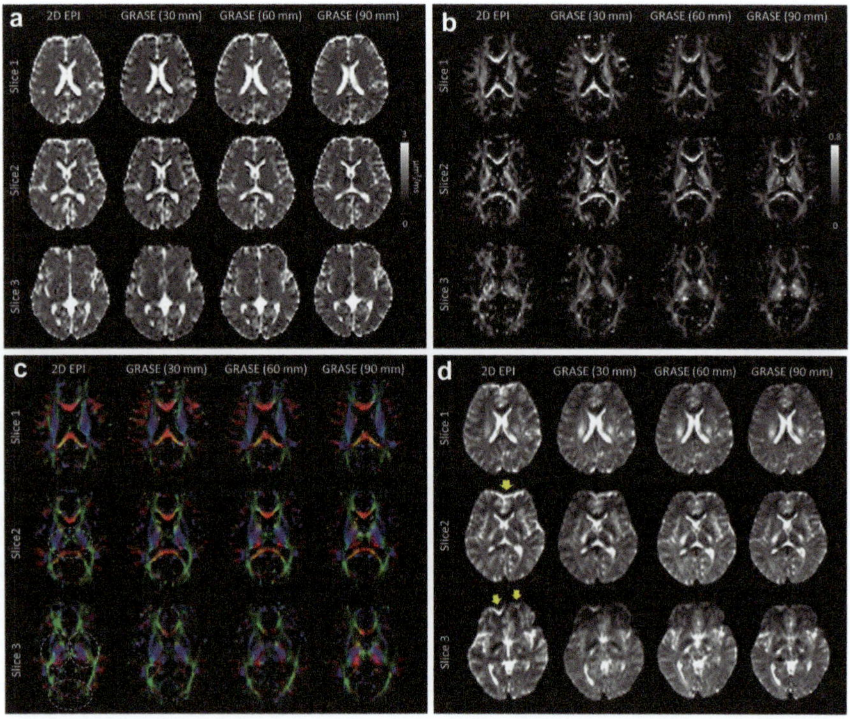

Fig. 3.6 Diffusion tensor imaging metrics measured by 3D OGprep-GRASE and 2D OG-EPI sequences. **a** ADC map. **b** FA map. **c** Direction-encoded color map. The dashed circles represent noisy areas in the images acquired by the EPI sequence, but the noise in these areas was improved by using the GRASE sequence and increasing the slab coverage thickness. **d** DWI map. The yellow arrow points to image distortion in the orbitofrontal cortex [26]

found in the white matter under the prefrontal cortex and the basal ganglia ($p < 0.001$). However, the results were quite close within the same type of measurement ($ADC_{50\,Hz} \approx ADC_{25\,Hz}$ 和 $ADC_{30\,ms} \approx ADC_{60\,ms}$).

3.5 Discussion

This study pioneeringly proposed the 3D OGprep-GRASE sequence, laying the foundation for the application of short diffusion time TDDMRI in clinical systems. This new sequence offers several advantages compared to traditional 2D multi-slice sequence imaging. Firstly, for TDDMRI based on 2D OG-EPI sequences, diffusion gradients need to be repeatedly applied to each slice, leading to a high duty cycle and long TR, which is particularly prominent in large amount slices imaging. In contrast, in 3D OGprep-GRASE sequence acquisition, the diffusion gradient is applied only once across the entire imaging volume, significantly reducing the total acquisition time. Specifically, in this experiment, the minimum TR for a 2D OG-EPI sequence with a 90 mm slab imaging range (30 slices) was 12,600 ms, while the minimum TR for the 3D OGprep-GRASE sequence was only 528 ms (excluding saturation recovery delay). Secondly, 3D acquisition can significantly improve SNR compared to 2D multi-slice imaging methods. For example, the image SNR was increased by 74% with a 60 mm slab imaging range and by 127% with a 90 mm slab imaging range. Therefore, the 3D OGprep-GRASE sequence yielded more accurate diffusion tensors with significantly reduced CV and noise in ADC and FA maps compared to the 2D OG-EPI sequence. Furthermore, this experiment also demonstrated that the 3D OGprep-GRASE sequence can effectively correct magnetic susceptibility artifacts in the 2D OG-EPI sequence, reducing geometric distortion in the images. The 3D GRASE sequence and the 2D EPI sequence showed similar sensitivity to the time-dependent changes in brain ADC.

A critical step in acquiring 3D *k*-space with multi-shot excitation is correcting phase errors caused by motion. Currently, several phase correction methods for multi-shot dMRI have been proposed, mainly categorized into navigator-based [27, 28] methods and reconstruction-based [29, 30] methods. In this study, a multiplexed sensitivity encoding [24] method was employed, which showed good performance in correcting phase errors in two-shot and three-shot excitation GRASE acquisitions. However, for more shots, navigators are needed to more effectively correct phase errors. Especially in 3D GRASE sequences, it is necessary to correct phase oscillations between odd and even spin echoes. Therefore, a twin-navigator strategy [31] is essential. Notably, phase errors in the OGprep-GRASE sequence are less pronounced compared to PGprep-GRASE, benefiting from the motion insensitivity and stronger eddy current resistance of the oscillating gradients themselves.

This study thoroughly explored the use of diffusion encoding gradients. They can be directly integrated into the readout module of spin-echo-type sequences or added as a preparation module before acquisition. This preparation strategy provides a highly flexible framework, allowing any type of diffusion encoding scheme to

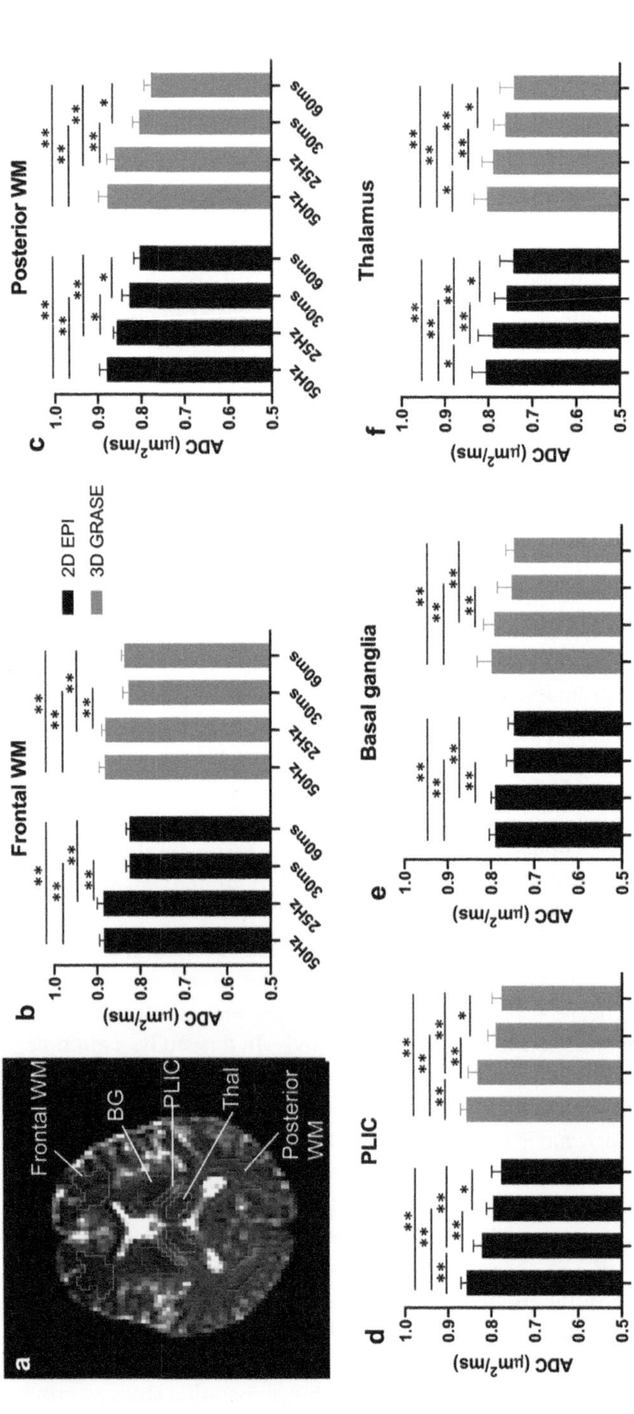

Fig. 3.7 Illustrates the diffusion time dependence of ADC measured by the 3D GRASE and 2D EPI sequences separately. **a** The defined ROIs, including frontal white matter (blue), posterior white matter (purple), posterior limb of the internal capsule (PLIC, green), basal ganglia (BG, orange), and thalamus (Thal, yellow). **b–f** The ADC in different regions at different diffusion times (OG = 50/25 Hz, PG = 30/60 ms) (*$p < 0.01$ and **$p < 0.001$ by ANOVA and post-hoc t-tests) [26]

be combined with readout trajectories (SSFP [18], MPRAGE [19], 3D EPI [17], etc.). This preparation strategy becomes particularly important when using oscillating gradients. Without this strategy, the echo spacing between subsequent 180° refocusing pulses would need to be long enough to accommodate the long oscillating gradients to match TE_1. However, because the $90°_{-x}$ excitation pulse in the diffusion preparation module removes phase information, phase errors at the end of diffusion encoding are converted into amplitude errors that are difficult to correct. Although gradient stabilizers [32] can correct these errors, they result in a 50% reduction in image SNR. Therefore, the SNR advantage of 3D acquisition is not obvious in 30 mm slab imaging. The SNR advantage of 3D acquisition over 2D acquisition only gradually increases as the slab imaging range increases to 60 or 90 mm, demonstrating the superiority and flexibility of the new sequence.

This study evaluated the diffusion time dependence effect of tissue scanned by the 3D GRASE sequence. TDdMRI, which can reveal the time-varying diffusion information of molecules, has achieved significant research results in preclinical and clinical studies. In preclinical research, it has been widely used in animal models of tumors [33-39] and brain diseases [7, 40], strongly promoting basic research on these diseases. In clinical research, its application mainly focuses on the early diagnosis of tumors and neurological diseases [41-46]. However, the 2D OG-EPI sequence used for TDDMRI has low SNR and resolution in dMRI images, making it difficult to diagnose small lesions in the brain. The 3D GRASE sequence proposed in this study not only significantly improved image SNR but also showed similar sensitivity to the time-dependent changes in brain ADC as the 2D EPI sequence. The 3D OGprep-GRASE sequence provides a solid foundation for subsequent research on TDDMRI techniques, contributing to improved imaging quality and efficiency, and promoting the gradual translation of TDDMRI technology from basic research to clinical applications.

Despite some significant achievements in this study, several limitations still exist. Firstly, the current imaging resolution (i.e., $2.75 \times 2.75 \times 3$ mm^3) is relatively low. The main purpose was to ensure that all scanning protocols were within a clinically acceptable time frame while maintaining a good SNR. Considering the scanning time limitations, these protocols are adequate for clinical use. Secondly, achieving 3D high-resolution whole-brain imaging requires the use of multi-shot acquisition techniques in the FSE phase encoding direction, which can lead to aliasing artifacts in the images. Therefore, navigators become crucial. Furthermore, the current 3D GRASE sequence is not yet equipped with fast imaging techniques, resulting in a relatively long readout time. Finally, the effects of CSF were not considered in low-resolution brain imaging. Subsequent chapters will gradually address these issues to enable TDDMRI to be truly used for clinical research.

3.6 Conclusion

The 3D OGprep-GRASE sequence proposed in this chapter on a 3 T clinical system significantly shortened the imaging time, improved image SNR, and reduced image distortion compared to the traditional 2D OG-EPI sequence. This enables reliable diffusion tensor reconstruction and the characterization of the diffusion time dependence of brain ADC using OG-dMRI. It is worth emphasizing that the results of this study give substantial prospects for the clinical application of OG-dMRI. The 3D OGprep-GRASE sequence is expected to promote the smooth transition of TDDMRI from the preclinical stage to clinical applications. In the future, these sequences will help provide more accurate and in-depth analysis tools for applications such as brain disease diagnosis and tracking disease progression.

References

1. Basser PJ, Mattiello J, Lebihan D (1994) MR diffusion tensor spectroscopy and imaging. Biophys J 66(1):259–267
2. Tuch DS, Reese TG, Wiegell MR et al (2002) High angular resolution diffusion imaging reveals intravoxel white matter fiber heterogeneity. Magnet Reson Med Off J Int Soc Magnet Reson Med 48(4):577–582
3. Wedeen VJ, Hagmann P, Tseng WYI et al (2005) Mapping complex tissue architecture with diffusion spectrum magnetic resonance imaging. Magnet Reson Med 54(6):1377–1386
4. Frank LR (2002) Characterization of anisotropy in high angular resolution diffusion-weighted MRI. Magnet Reson Med Off J Int Soc Magnet Reson Med 47(6):1083–1099
5. Westin C-F, Knutsson H, Pasternak O et al (2016) Q-space trajectory imaging for multidimensional diffusion MRI of the human brain. Neuroimage 135:345–362
6. Gore JC, Xu J, Colvin DC et al (2010) Characterization of tissue structure at varying length scales using temporal diffusion spectroscopy. NMR Biomed 23(7):745–756
7. Novikov DS, Jensen JH, Helpern JA et al (2014) Revealing mesoscopic structural universality with diffusion. Proceed Natl Acad Sci 111(14):5088–5093
8. Does MD, Parsons EC, Gore JC (2003) Oscillating gradient measurements of water diffusion in normal and globally ischemic rat brain. Magnet Reson Med Off J Int Soc Magnet Reson Med 49(2):206–215
9. Veraart J, Novikov DS, Fieremans E (2018) TE dependent diffusion imaging (TEdDI) distinguishes between compartmental T2 relaxation times. Neuroimage 182:360–369
10. Hutter J, Slator PJ, Christiaens D et al (2018) Integrated and efficient diffusion-relaxometry using ZEBRA. Sci Rep 8(1):15138
11. Reynaud O (2017) Time-dependent diffusion MRI in cancer: tissue modeling and applications. Front Phys 5:58
12. Novikov DS, Fieremans E, Jespersen SN et al (2019) Quantifying brain microstructure with diffusion MRI: theory and parameter estimation. NMR Biomed 32(4):e3998
13. Aggarwal M, Burnsed J, Martin LJ et al (2014) Imaging neurodegeneration in the mouse hippocampus after neonatal hypoxia–ischemia using oscillating gradient diffusion MRI. Magnet Reson Med 72(3):829–840
14. Pyatigorskaya N, Le Bihan D, Reynaud O et al (2014) Relationship between the diffusion time and the diffusion MRI signal observed at 17.2 Tesla in the healthy rat brain cortex. Magnet Reson Med 72(2):492–500

15. Wu D, Martin LJ, Northington FJ et al (2014) Oscillating gradient diffusion MRI reveals unique microstructural information in normal and hypoxia-ischemia injured mouse brains. Magnet Reson Med 72(5):1366–1374
16. Jiang X, Mckinley ET, Xie J et al (2019) In vivo magnetic resonance imaging of treatment-induced apoptosis. Sci Rep 9(1):9540
17. Moeller S, Ramanna S, Lenglet C et al (2020) Self-navigation for 3D multishot EPI with data-reference. Magnet Reson Med 84(4):1747–1762
18. Ohalloran R, Aksoy M, Van A et al (2013) 3D isotropic high-resolution diffusion-weighted MRI of the whole brain with a motion-corrected steady-state free precession sequence. Magnet Reson Med 70(2):466–478
19. Hikishima K, Yagi K, Numano T et al (2008) Volumetric q-space imaging by 3D diffusion-weighted MRI. Magnet Reson Imag 26(4):437–445
20. Aggarwal M, Mori S, Shimogori T et al (2010) Three-dimensional diffusion tensor microimaging for anatomical characterization of the mouse brain. Magnet Reson Med 64(1):249–261
21. Piredda GF, Hilbert T, Canales-Rodríguez EJ et al (2021) Fast and high-resolution myelin water imaging: accelerating multi-echo GRASE with CAIPIRINHA. Magnet Reson Med 85(1):209–222
22. Yin M, Cao G, Lv S et al (2024) Intravoxel incoherent motion diffusion-weighted imaging of solitary pulmonary lesions: initial study with gradient-and spin-echo sequences. Clin Radiol 32:546
23. Van AT, Cervantes B, Kooijman H et al (2017) Analysis of phase error effects in multishot diffusion-prepared turbo spin echo imaging. Quant Imag Med Surg 7(2):238
24. Chen N-K, Guidon A, Chang H-C et al (2013) A robust multi-shot scan strategy for high-resolution diffusion weighted MRI enabled by multiplexed sensitivity-encoding (MUSE). Neuroimage 72:41–47
25. Farrell JA, Landman BA, Jones CK et al (2007) Effects of signal-to-noise ratio on the accuracy and reproducibility of diffusion tensor imaging–derived fractional anisotropy, mean diffusivity, and principal eigenvector measurements at 1.5 T. J Magnet Reson Imag Off J Int Soc Magnet Reson Med 26(3):756–767
26. Wu D, Liu D, Hsu YC et al (2021) Diffusion-prepared 3D gradient spin-echo sequence for improved oscillating gradient diffusion MRI [J]. Magnet reson med 85(1):78–88
27. Atkinson D, Counsell S, Hajnal JV et al (2006) Nonlinear phase correction of navigated multi-coil diffusion images. Magnet Reson Med Off J Int Soc Magnet Reson Med 56(5):1135–1139
28. Miller KL, Pauly JM (2003) Nonlinear phase correction for navigated diffusion imaging. Magnet Reson Med Off J Int Soc Magnet Reson Med 50(2):343–353
29. Guhaniyogi S, Chu ML, Chang HC et al (2016) Motion immune diffusion imaging using augmented MUSE for high-resolution multi-shot EPI. Magnet Reson Med 75(2):639–652
30. Mani M, Jacob M, Kelley D et al (2017) Multi-shot sensitivity-encoded diffusion data recovery using structured low-rank matrix completion (MUSSELS). Magnet Reson Med 78(2):494–507
31. Mori S, Van Zijl PC (1998) A motion correction scheme by twin-echo navigation for diffusion-weighted magnetic resonance imaging with multiple RF echo acquisition. Magnet Reson Med 40(4):511–516
32. Jeong EK, Kim SE, Kholmovski EG et al (2006) High-resolution DTI of a localized volume using 3D single-shot diffusion-weighted ST imulated echo-planar imaging (3D ss-DWSTEPI). Magnet Reson Med Off J Int Soc Magnet Reson Med 56(6):1173–1181
33. Jiang X, Li H, Xie J et al (2017) In vivo imaging of cancer cell size and cellularity using temporal diffusion spectroscopy. Magnet Reson Med 78(1):156–164
34. Jiang X, Xu J, Gore JC (2019) Quantitative temporal diffusion spectroscopy as an early imaging biomarker of radiation therapeutic response in gliomas: a preclinical proof of concept. Adv Radiat Oncol 4(2):367–376
35. Xiaoyu J, Stephanie D, Kathryn EB et al (2020) MRI of tumor T cell infiltration in response to checkpoint inhibitor therapy. J ImmunoTherapy Cancer 8(1):e000328

36. Jiang X, Mckinley ET, Xie J et al (2024) Detection of treatment response in triple-negative breast tumors to paclitaxel using MRI cell size imaging. J Magnet Reson Imag 59(2):575–584
37. Reynaud O, Winters KV, Hoang DM et al (2016) Pulsed and oscillating gradient MRI for assessment of cell size and extracellular space (POMACE) in mouse gliomas. NMR Biomed 29(10):1350–1363
38. Panagiotaki E, Walker-Samuel S, Siow B et al (2014) Noninvasive quantification of solid tumor microstructure using VERDICT MRI. Cancer Res 74(7):1902–1912
39. Jiang X, Li H, Xie J et al (2016) Quantification of cell size using temporal diffusion spectroscopy. Magnet Reson Med 75(3):1076–1085
40. Lee H-H, Papaioannou A, Novikov DS et al (2020) In vivo observation and biophysical interpretation of time-dependent diffusion in human cortical gray matter. Neuroimage 222:117054
41. Zhang J, Lemberskiy G, Moy L et al (2021) Measurement of cellular-interstitial water exchange time in tumors based on diffusion-time-dependent diffusional kurtosis imaging. NMR Biomed 34(6):e4496
42. Solomon E, Lemberskiy G, Baete S et al (2023) Time-dependent diffusivity and kurtosis in phantoms and patients with head and neck cancer. Magnet Reson Med 89(2):522–535
43. Xu J, Jiang X, Li H et al (2020) Magnetic resonance imaging of mean cell size in human breast tumors. Magnet Reson Med 83(6):2002–2014
44. Panagiotaki E, Chan RW, Dikaios N et al (2015) Microstructural characterization of normal and malignant human prostate tissue with vascular, extracellular, and restricted diffusion for cytometry in tumours magnetic resonance imaging. Invest Radiol 50(4):548
45. Lemberskiy G, Feiweier T, Gyftopoulos S et al (2021) Assessment of myofiber microstructure changes due to atrophy and recovery with time-dependent diffusion MRI. NMR Biomed 34(7):e4534
46. Fieremans E, Lemberskiy G, Veraart J et al (2017) In vivo measurement of membrane permeability and myofiber size in human muscle using time-dependent diffusion tensor imaging and the random permeable barrier model. NMR Biomed 30(3):e3612

Chapter 4
Twin-Navigator based 3D Gradient and Spin Echo Imaging

Abstract This chapter introduces twin-navigator-based multi-shot acquisition 3D oscillating gradient prepared gradient and spin echo (3D OGprep-GRASE) and pulsed gradient prepared gradient and spin echo (3D PGprep-GRASE) sequences for whole-brain time-dependent diffusion MRI (TDDMRI) on a clinical 3 T system. Building upon the single-shot sequences from Chap. 3, this development aims to overcome the limitation of obtaining whole-brain data. The newly proposed sequences incorporate twin 1D and 2D navigator acquisition schemes to effectively correct phase errors inherent in multi-shot acquisitions. Experimental results demonstrated that navigator-based phase correction significantly reduced artifacts in DWI images. Compared to conventional 2D EPI sequences, the 3D GRASE sequences showed significantly higher image signal-to-noise ratio (SNR). Furthermore, the study revealed that the 3D sequences could discover more abundant cortical fibers and a higher fiber bundle cross-sectional area in the corticospinal tract in fiber tracking and microstructure parameter estimation compared to 2D EPI. These findings highlight the potential of the proposed 3D GRASE sequences for advanced whole-brain TDDMRI research and neuroscience applications.

Keywords Twin-navigator · Multi-shot acquisition · Phase correction · Fiber tracking · SNR · GRAPPA · Microstructure

4.1 Introduction

According to the contents of Sect. 1.4 in Chap. 1, 3D diffusion magnetic resonance imaging (dMRI) sequences are highly favored by researchers due to their characteristics of high resolution and high signal noise ratio (SNR) efficiency, especially when using 3D *k*-space for encoding thin slices, which has significant advantages. Therefore, 3D acquisition is undoubtedly the ideal choice for high-resolution time-dependent (TD) dMRI imaging at short diffusion times, as it can effectively compensate for the loss of image SNR caused by long diffusion-weighted attenuation and can also detect tissue microstructure and fiber bundle connections in continuous 3D

D. Wu et al., *Development and Application of 3D Diffusion Magnetic Resonance Imaging Sequences*, Advanced Topics in Science and Technology in China 5,
https://doi.org/10.1007/978-981-95-0808-2_4

space. Among these, 3D GRASE acquisition integrates FSE readout in one phase-encoding dimension and EPI readout in another. This strategy equips it with the high SNR efficiency of 3D encoding and low magnetic susceptibility artifacts. In addition, some pseudo-3D acquisition strategies have been proposed, including multi-slab [1–4] and generalized slice dithered enhanced resolution [5, 6], attempting to bypass the lengthy phase-encoding steps in 3D *k*-space. However, the residual slab boundary artifacts that occur with these pseudo-3D acquisition methods are a complex and critical problem that needs to be solved. In practical applications, multi-shot acquisition of 3D *k*-space is particularly necessary to truly achieve 3D TDDMRI whole-brain acquisition. However, multi-shot acquisition diffusion-encoded imaging faces challenges from eddy current and motion-induced phase errors [7, 8]. Therefore, phase correction must be performed to eliminate these errors before merging the *k*-space segments obtained from multi-shot acquisition.

The 3D OGprep-GRASE and PGprep-GRASE sequences developed in Chap. 3 acquired image data through single-shot acquisition [9], achieving certain effectiveness in studying the application of TDDMRI imaging in brain diffusion time dependence. However, this study failed to obtain whole-brain data. To acquire whole-brain TDDMRI data and ensure quality, this study proposed multi-shot acquisition of 3D OGprep-GRASE and PGprep-GRASE sequences on its basis, equipped with twin 1D or twin 2D navigator acquisition schemes. To test the phase calibration capability of the new sequences, relevant experiments were conducted on a Siemens magnetic resonance machine. TDDMRI whole-brain image data is clinically commonly used for fiber tracking and microstructure parameter estimation. However, their common practice is based on multi-slice image acquisition using 2D EPI sequences. Although nominally the slice thickness and in-plane resolution of 2D sequence acquisition are the same, this sequence acquisition method introduces deviations in fiber tracking in the in-plane and perpendicular plane directions. This problem has been proposed and discussed in the academic community, but no specific quantitative evaluation results and solutions have been obtained yet [10]. Therefore, this study performed whole-brain TDDMRI imaging using the newly proposed 3D sequences and 2D sequences, and quantitatively evaluated the results of fiber tracking and microstructure parameter estimation using DTI and higher-order fiber orientation density (FOD) based metrics [11], to deeply analyze the specific effects of different sequences on fiber tracking and microstructure.

4.2 Theory

4.2.1 The Twin-Navigator Acquisition Module

The twin-navigator-based multi-shot acquisition 3D OGprep-GRASE sequence mainly consists of five modules, as shown in Fig. 4.1, including: a global stabilization module, a diffusion preparation module, a fat saturation module, a 3D

GRASE acquisition module, and a twin-navigator acquisition module. Meanwhile, a twin-navigator-based multi-shot acquisition 3D PGprep-GRASE sequence was also designed in this chapter.

The 3D GRASE readout module consists of linear EPI encoding and central FSE encoding, achieving GRAPPA acceleration in the EPI phase-encoding direction, and dividing the FSE encoding into segments acquired in an interleaved manner for multi-shot *k*-space acquisition. The diagram and pseudo-code of the *k*-space encoding trajectory in the *ky*–*kz* plane are shown in Figs. 4.2 and 4.3.

The twin-navigator acquisition module is embedded at the end of the sequence and repeated twice, specifically to correct the phase errors of even and odd spin echoes. The twin 2D navigator performs phase encoding along the EPI direction but

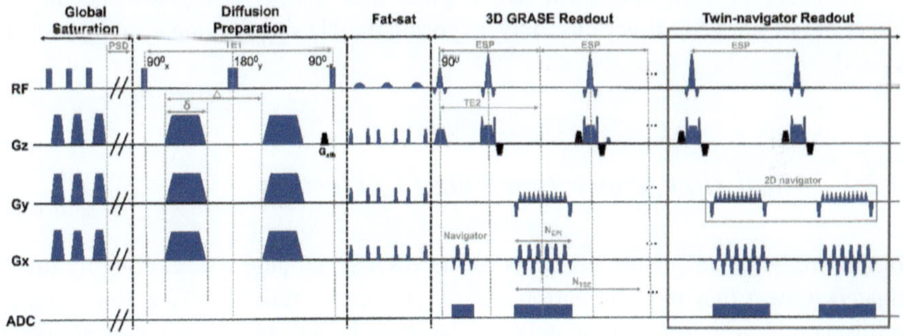

Fig. 4.1 The schematic of the twin-navigator-based multi-shot acquisition 3D OGprep-GRASE sequence [29]

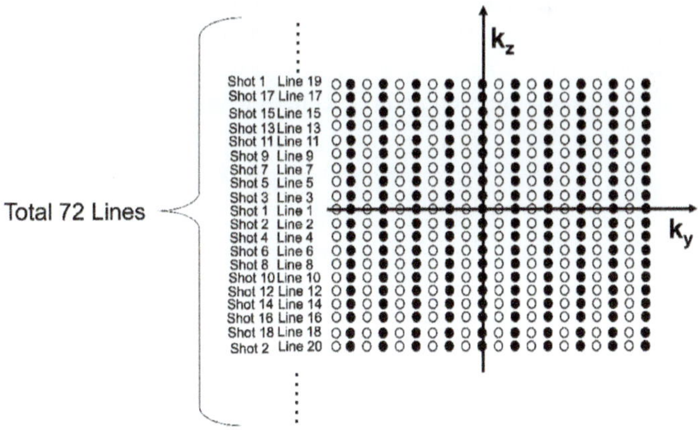

Fig. 4.2 The phase-encoding trajectory of the multi-shot acquisition 3D GRASE readout module in the $k_z - k_y$ space. The k_x direction is fully sampled, and a twofold GRAPPA acceleration is applied in k_y. Solid and hollow circles represent acquired and unacquired *k*-space data for each shot, respectively. A total of 72 k_z lines are acquired with 18 shots, and each shot employs a central encoding $N_{\mathrm{FSE}} = 4$

Fig. 4.3 The pseudo-code for the $k_y - k_z$ encoding trajectory of the multi-shot acquisition 3D GRASE readout module

```
Algorithm K-space ordering
for k = 1:4
# each shot has 4 TSE acquisition named TSE k
    for m =1:18
# each K-space excitation named "Shot m"
        n = 18×k+m-1;
# each Kz line named Line n
end
```

not along the FSE direction, acquiring the 2D signal of the central k-space plane at $k_z = 0$ with N_{EPI} lines. The twin 1D navigator does not perform phase encoding in either the EPI or FSE directions, only acquiring the 1D signal along k_x at $k_y = 0$ and $k_z = 0$.

4.2.2 *Multi-shot Acquisition Phase Correction*

In this study, the acquisition process of the GRASE module was thoroughly investigated. Given that the acquisition time of the echo train is only a few hundred milliseconds, which is significantly shorter than the TR, phase errors do not occur during the acquisition within the same shot but rather across different shots. For the twin 2D navigator, phase encoding was performed along the EPI direction, and navigator signals were acquired. Subsequently, phase error correction was conducted in the hybrid x–y–k_z space. This involved first aligning the phases between the even and odd navigators, $\varphi_{n,o}(x, y, 0)$ and $\varphi_{n,e}(x, y, 0)$ and their corresponding even and odd image echoes $\psi_{i,o}(x, y, kz_o)$ and $\psi_{i,e}(x, y, kz_e)$ within each shot, and correcting the phase errors between the even and odd echoes using the following equations:

$$\psi'_{i,o}(x, y, kz_o) = \psi_{i,o}(x, y, kz_o) - \varphi_{n,o}(x, y, 0) \tag{4.1}$$

$$\psi'_{i,e}(x, y, kz_e) = \psi_{i,e}(x, y, kz_e) - \varphi_{n,e}(x, y, 0) \tag{4.2}$$

where "o" and "e" denote the odd and even echoes, respectively. Then, the even and odd navigators were utilized to correct the phase errors of the corresponding even and odd image echoes between different shots. Let $\varphi_{n,o,s}(x, y, 0)$ and $\varphi_{n,e,s}(x, y, 0)$ be the navigator echo signals of the odd and even echoes for the s-th shot, and $\psi_{i,o,s}(x, y, kz_o)$ and $\psi_{i,e,s}(x, y, kz_o)$ be the image echo signals corresponding to the odd and even echoes for the s-th shot. The final corrected phases for the even and odd images $\psi''_{i,o,s}(x, y, kz_o)$ and $\psi''_{i,e,s}(x, y, kz_e)$ were calculated as:

$$\psi''_{i,o,s}(x, y, kz_o) = \psi'_{i,o,s}(x, y, kz_o) - \varphi_{n,o,s}(x, y, 0) \tag{4.3}$$

$$\psi''_{i,e,s}(x, y, kz_e) = \psi'_{i,e,s}(x, y, kz_e) - \varphi_{n,e,s}(x, y, 0) \tag{4.4}$$

For the twin 1D navigator, navigator signals were acquired without any phase encoding, and phase error correction was performed line by line in the hybrid x–k_y–k_z space. This process was consistent with the phase correction procedure based on the twin 2D navigator.

4.2.3 2D GRAPPA Accelerated Imaging

GRAPPA is a parallel acquisition technique that improves the speed of dMRI image acquisition and is now widely applied in sequence acquisition. It utilizes receiver coils to simultaneously acquire partial k-space data, and based on the relationships of each coil in k-space, it down samples the k-space, significantly improving acquisition efficiency. In this study, we also employed this technique in the design of the 3D GRASE readout module. This module consists of EPI encoding with linear encoding and FSE encoding with central encoding, cleverly integrating two distinct sequence encoding methods. On this basis, GRAPPA technology was applied in the EPI phase encoding direction.

4.3 Experimental Methods

4.3.1 Magnetic Resonance Experiment

This study utilized in-house developed twin-navigator-based multi-shot acquisition 3D OGprep-GRASE and 3D PGprep-GRASE sequences, as well as a 2D EPI sequence, for data acquisition. A total of seven healthy volunteers (average age = 23 ± 1 years, range 20–25 years, all male) were recruited for this study. All scans were performed on a Siemens 3 T MAGNETOM Prisma magnetic resonance scanner equipped with a 64-channel head receiving coil. The scanner's maximum gradient was 80 mT/m, and the maximum slew rate was 200 mT/m/ms. This study established three important objectives: (1) to investigate the impact of adding the proposed twin 1D or twin 2D navigator on image quality; (2) to compare the difference in image SNR between the proposed sequences and the 2D EPI sequence; and (3) to compare the differences in fiber tracking and microstructure parameter estimation between the proposed sequences and the 2D EPI sequence. Based on these objectives, two targeted experimental schemes were designed. Prior to the experiments, all volunteers provided informed consent after fully understanding the study's purpose and procedures. All research procedures were approved by the institutional ethics review board.

4.3.2 Data Acquisition

Experiment 1: To investigate the effect of adding a twin 1D or twin 2D navigator to the 3D GRASE sequence on image quality and to compare the SNR of the 3D GRASE and 2D EPI sequences at the same scan time. The 3D OGprep-GRASE sequence parameters were: TR/TE1/TE2 = 2500/103.46/46 ms, $N_{\mathrm{EPI}} = 67$, $N_{\mathrm{FSE}} = 4$, with 18 excitations acquired in the spin echo encoding direction. The readout matrix size of the navigation module was the same as the imaging matrix (146 × 134), and the readout time was 100 ms. Data were acquired using a 2D OG-EPI sequence with TR/TE = 11,250/150 ms, $N_{\mathrm{EPI}} = 110$. Other parameters were kept the same in both sequences: oscillation frequency = 20 Hz (1 cycle, $\Delta_{\mathrm{eff}} \approx 12.5$ ms), FOV = 220 × 200 × 144 mm^3, resolution = 1.5 × 1.5 × 2 mm^3, b = 1000 s/mm^2, 30 diffusion directions and 4 b_0 images were acquired for each b value, and the GRAPPA acceleration factor was set to 2. The 3D sequence was acquired once, and the 2D sequence was repeated 4 times. The 3D PGprep-GRASE and 2D PG-EPI sequences had the same acquisition parameters as the above sequences, except for TE = 83 ms.

Experiment 2: To compare the differences between the 3D GRASE and 2D EPI sequences in fiber tracking and microstructure parameter estimation. Based on the results of Experiment 1, the 2D EPI was repeated 9 times in this experiment to match the image SNR of the single acquisition of the 3D GRASE. All other imaging parameter settings were the same as in Experiment 1.

4.3.3 Data Analysis

The k-space data acquired by the 3D GRASE sequence were reconstructed using Tikhonov-regularized GRAPPA [12] and twin-navigator-based data phase correction. Subsequently, preprocessing steps including Gibbs artifact removal, eddy current correction, image registration, and bias field correction were performed using MRtrix software.

On the 3D GRASE and 2D EPI data from Experiment 1, the SNR was calculated in ROIs such as the white matter, basal ganglia, and thalamus, as shown in Fig. 4.5. In Experiment 2, whole-brain fiber tracking was performed using a probabilistic algorithm based on FOD in MRtrix. The fiber tracking imaging parameters were: number of required fibers = 100,000, cutoff FOD = 0.15, and other parameters were default values (step size = 0.2 mm, angle = 60°, minimum fiber length = 10 mm, maximum fiber length = 200 mm). In this study, fiber morphology was quantified by mapping fibers to the image, and the parameters included: fiber length and volume. This study also obtained fiber microstructure indices through a fixel based analysis (FBA) [11] pipeline, such as fibre density (FD), fibre cross-section (FC), and fibre density and cross-section (FDC).

Analysis of variance (ANOVA) was used to evaluate the difference in SNR between the two sequence images in different ROIs, and then post-hoc *t*-tests were performed in MATLAB on the SNR obtained from the two sequences in each ROI. Using the statistical tools in MRtrix, a two-sample paired *t*-test was conducted using the threshold free cluster enhancement (TFCE) method [11] within the white matter region to compare the FBA-based indices FD, FC, and FDC between the 2D EPI and 3D GRASE sequences. Subsequently, the fixel space was transformed into volume space, and ANOVA and post-hoc *t*-tests were used to compare the fiber morphology indices of the regions with statistically significant differences in the above analysis: fiber length and volume. For all results, a Bonferroni-corrected $P < 0.05$ was considered statistically significant.

4.4 Experimental Results

The results of computer-simulated PSF with GRAPPA technology applied in the EPI phase encoding direction of the GRASE sequence are shown in Fig. 4.4. GRAPPA technology not only reduced TE and shortened acquisition time but also improved the PSF profile, increased the effective spatial resolution, and provided more accurate data sources for subsequent analysis.

Since PG-weighted sequences are more susceptible to aliasing artifacts, Fig. 4.5 shows DWI images of the twin-navigator-based 3D PGprep-GRASE sequence obtained in axial, coronal, and sagittal views. These images correspond to two *b*-values, 0 and 1000 s/mm^2. This study compared these images before and after twin 1D and 2D navigator phase correction. The experimental results showed that without navigator phase correction, although the b_0 image showed robustness in the reconstruction of multi-shot 3D GRASE images, directly reconstructed DWI images had severe phase errors and resulting image artifacts in all diffusion directions. However,

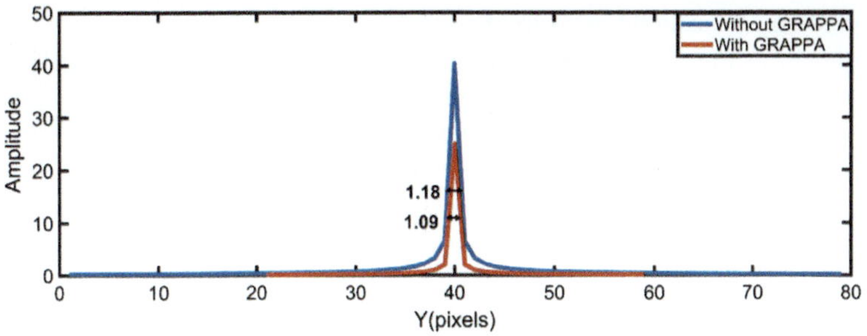

Fig. 4.4 PSF of the 3D OGprep-GRASE sequence in the EPI phase encoding direction. Includes cases with GRAPPA acceleration (orange line) and without GRAPPA acceleration (blue line). In the sequence with GRAPPA acceleration, the full width at half maximum (FWHM) of the PSF was 1.09, while it was 1.18 in the sequence without GRAPPA acceleration. Where TE = 150 ms, assuming T2 and T2* of brain tissue were 80 and 50 ms, respectively

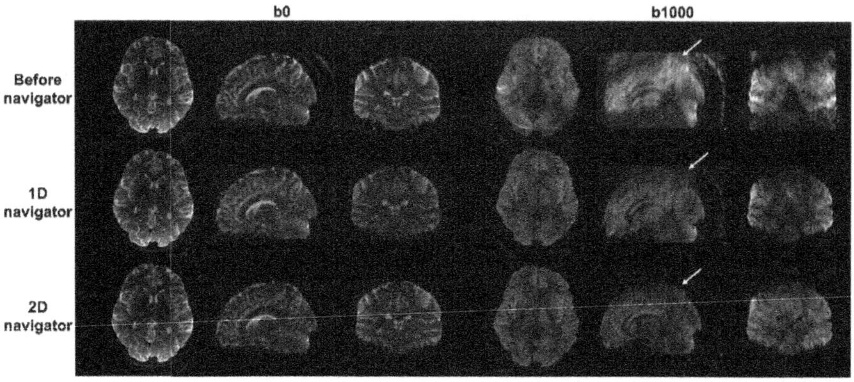

Fig. 4.5 Images before and after twin 1D and twin 2D navigator phase correction. Axial, sagittal, and coronal views of 3D GRASE and 2D EPI sequences acquired with *b*-values of 0 and 1000 s/mm^2 before and after correction with twin 1D or twin 2D navigators

after phase correction with the navigators, the phase errors in the DWI images were significantly reduced. Further comparison revealed that the phase error correction effect of the twin 2D navigator was superior to that of the twin 1D navigator. For example, the twin 2D navigator could more effectively eliminate residual artifacts along the *z*-direction (white arrows).

As shown in Fig. 4.6, this experiment quantitatively compared the SNR of images scanned by the 2D EPI and 3D GRASE sequences. It is evident that in the ROIs of white matter (WM), basal ganglia (BG), and thalamus (Thal), for both b_0 and b_{1000} images, the SNR of images scanned by the 3D OGprep-GRASE sequence was significantly higher than that of the 2D OG-EPI sequence, with increases of 25, 44, and 52% respectively ($p < 0.01$) (Fig. 4.6b). In addition, the SNR of images from the 3D PGprep-GRASE sequence was higher than that of the 2D PG-EPI sequence by 43, 46, and 41% respectively ($p < 0.01$) (Fig. 4.6c). The above results indicate that the twin-navigator-based 3D GRASE sequence has advantages in image SNR and can provide more accurate measurement results.

As shown in Fig. 4.7, this study utilized DWI data scanned by 2D EPI and 3D GRASE sequences to generate whole-brain probabilistic fiber tracking maps and colormaps based on the DTI model. These images vividly illustrate the distribution of nerve fibers in various brain regions. Notably, the nerve fibers shown in the images obtained by the two sequences exhibited significant differences in some cortical regions and deep gray matter nuclei of the brain. The data scanned by the 3D sequence could better connect the subcortical white matter regions to the cortical gyri. This suggests that the 3D sequence has a potential advantage in tracing complex U-shaped nerve fibers, while the 2D data was relatively inferior in this aspect. Furthermore, the colormap constructed from the 3D data showed higher fiber anisotropy and more microstructural details in the cerebral cortex regions, whereas the display effect of the colormap in the cortical regions from the 2D data was not ideal. This finding is consistent with previous research reports on 3D sequences [13], further confirming

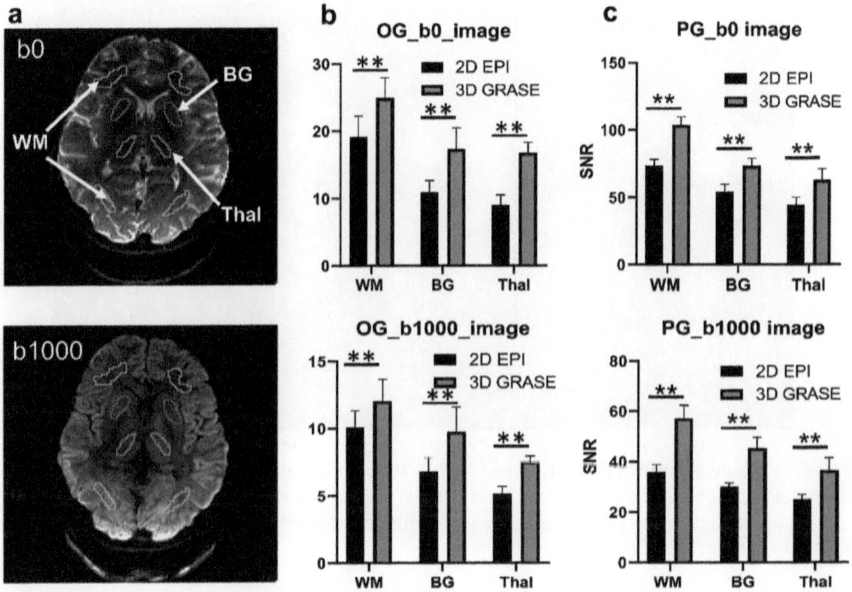

Fig. 4.6 SNR of images scanned by 3D GRASE and 2D EPI sequences. **a** ROIs for image comparison at *b*-values of 0 and 1000 s/mm^2 include white matter (WM, blue), basal ganglia (BG, pink), and thalamus (Thal, green). In all ROIs and *b*-values, the results show that the SNR of the 3D OGprep-GRASE sequence is significantly ($p < 0.01$) higher than that of the 2D OG-EPI sequence (**b**) and the SNR of the 3D PGprep-GRASE sequence is significantly ($p < 0.01$) higher than that of the 2D PG-EPI sequence (**c**)

the importance and superiority of 3D sequences in image reconstruction and data analysis.

As shown in Fig. 4.8a, this experiment performed a whole-brain FBA pipeline to quantitatively compare microstructural parameters between the 2D OG-EPI and 3D OGprep-GRASE sequences. It was found that the 3D sequence had a higher fibre cross-section in the longitudinally extending (i.e., inter-layer direction) corticospinal tract (CST), with the difference ranging from 0 to 20%. This finding further confirmed that using the 3D sequence could more accurately obtain the microstructural parameters of brain fibers. In addition, this experiment further compared the fiber morphology indices of the corticospinal tract obtained from 2 and 3D data. The results, as shown in Fig. 4.8b, indicated that the fiber length in the corticospinal tract from the 3D data was significantly shorter, and the fiber volume was significantly larger ($p < 0.05$). This suggests that the 3D sequence can capture fiber bundles with larger cross-sectional areas, which is consistent with the results obtained by FBA. Furthermore, this experiment performed the same analysis on the 2D PG-EPI and 3D PGprep-GRASE sequences and obtained completely consistent results with the OG sequences, except that the difference range of the fibre cross-section was 0 ~ 10%, as shown in Fig. 4.9.

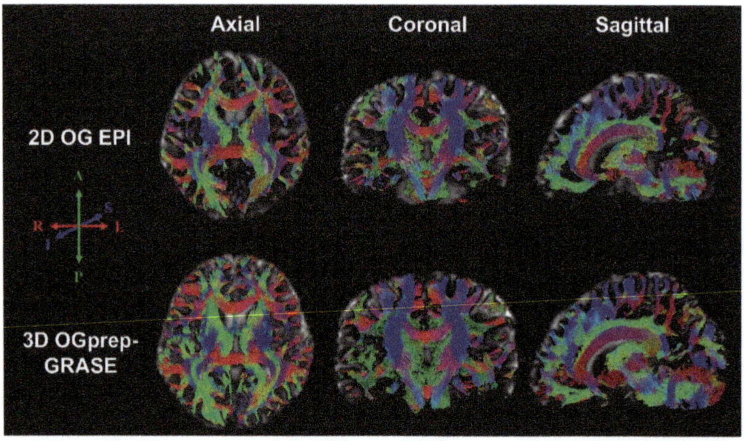

Fig. 4.7 Whole-brain fiber tracking obtained with 3D OGprep-GRASE and 2D OG-EPI sequences. Images are divided into axial, sagittal, and coronal views. Color coding indicates local direction (red represents left–right direction, green represents anterior–posterior direction, blue represents superior-inferior direction)

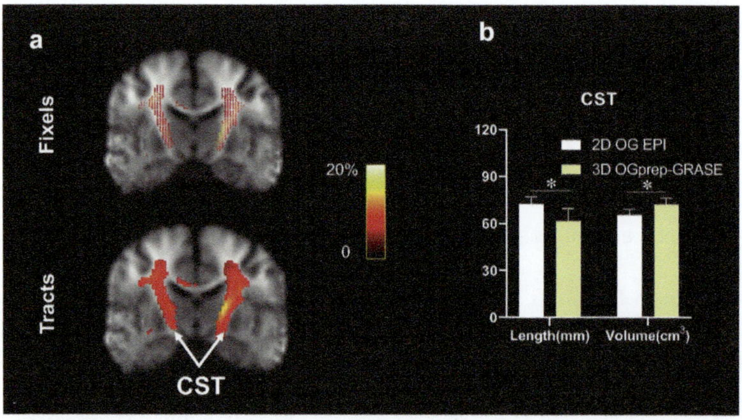

Fig. 4.8 Quantitative comparison of microstructural parameters between 3D OGprep-GRASE and 2D OG-EPI sequences. **a** Significant differences in fibre cross-section between the two sequences ($p < 0.05$), with results plotted along the corresponding fiber bundles. Fixels and fibers are color-coded by percentage difference (3D–2D) and overlaid on the FOD template map. **b** Comparison of morphological parameters along the corticospinal tract, including fiber length and volume

4.5 Discussion

The goal of this study was to develop twin navigator-based multi-shot acquisition 3D OGprep-GRASE and PGprep-GRASE sequences to achieve whole-brain TDDMRI acquisition on a clinical 3 T system. This study compared the performance of the

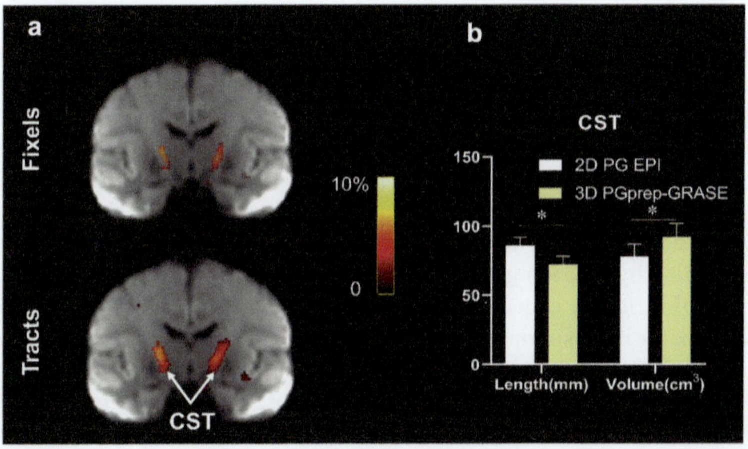

Fig. 4.9 Quantitative comparison of microstructural parameters between 3D PGprep-GRASE and 2D PG-EPI sequences. **a** Significant differences in fibre cross-section between the two sequences ($p < 0.05$), with results plotted along the corresponding fiber bundles. Fixels and fibers are color-coded by percentage difference (3D–2D) and overlaid on the FOD template map. **b** Comparison of morphological parameters along the corticospinal tract, including fiber length and volume [29]

3D GRASE sequence and the conventional 2D EPI sequence in aspects such as fiber tracking and microstructural parameter estimation. The twin navigator can accurately correct the phase errors of the 3D GRASE sequence in multi-shot acquisition, thereby realizing whole-brain 3D TDDMRI acquisition. Compared to images from the 2D EPI sequence with the same scan time, the 3D GRASE sequence had significantly higher image SNR. Meanwhile, the 3D GRASE sequence also employed 2D GRAPPA parallel acquisition technique to shorten TE and improve the PSF profile [14]. Furthermore, through quantitative comparison of fiber tracking results in the cerebral cortex regions and the microstructural parameters of fibers in the corticospinal tract obtained by 2D and 3D sequences, significant differences were found in the results obtained by the two sequences. These research results indicate that this study provides strong technical support for realizing whole-brain TDDMRI scans on clinical 3 T MRI systems, and the proposed sequences can obtain more accurate fiber microstructural parameters.

3D dMRI sequences have shown unique advantages in improving image SNR and optimizing slice profiles [7, 15, 16]. GRASE encoding combines the advantages of FSE and EPI encoding, realizing the reduction of susceptibility artifacts while rapidly acquiring images, making it an ideal choice for balancing acquisition speed and image artifacts. In previous related studies, Manjunathan et al. compared the performance of different pseudo-continuous arterial spin labeling sequences (2D simultaneous multi-slice readout EPI, 2D EPI, and 3D GRASE sequence combined with background suppression module) in perfusion MRI studies. Their results showed that the 3D GRASE sequence exhibited the lowest cerebral blood flow variability among all methods, indicating that the 3D GRASE sequence has the highest robustness to noise. Shao et al. [17] combined a diffusion preparation module, pseudo-continuous arterial spin labeling, and 3D GRASE encoding to quantitatively measure the water exchange rate of the blood–brain barrier. However, this study only used a single-shot acquisition GRASE sequence, making it difficult to achieve high-resolution whole-brain acquisition. Our study proposed the 3D OGprep-GRASE sequence in Chap. 3, which improved image SNR and shortened scan time on a clinical MRI scanner, but still faced the same problem. To solve this issue, this chapter introduced the multi-shot acquisition technique, achieving the goal of higher resolution whole-brain imaging by extending the scan time. However, it should be noted that in the presence of diffusion encoding, the multi-shot acquisition readout technique can introduce phase inconsistency issues that require appropriate measures for processing and optimization.

Numerous innovative schemes have been proposed for merging multi-shot acquisition data and eliminating phase errors. Porter and Heidemann [18] constructed a segmented readout EPI sequence and a 2D navigator-based acquisition method to correct phase errors. However, this method requires long scan times for each imaging volume, leading to the introduction of additional motion artifacts. In addition, Moeller et al. [19] proposed a self-navigator strategy aimed at improving the reconstruction of diffusion-weighted 3D multi-shot acquisition EPI. However, this new method for correcting phase errors heavily relies on the slice profile of the excitation slab. This is particularly problematic because clinical MRI machines typically cannot produce perfect RF pulses. To address these issues, this study employed a twin navigator technique [20]. This technique can track the phase of odd and even echoes and correct intra-echo and inter-echo phase errors. This technique has shown significant advantages in multi-shot acquisition 3D GRASE imaging of the mouse brain [21, 22]. However, considering that the test results of this technique on clinical scanning equipment are yet to be observed, and the differences between mice and human subjects are significant (mice are fixed and anesthetized, with relatively less motion), it is necessary to re-evaluate the potential of this technique for human brain applications. This is crucial for the clinical application of twin navigator techniques. To address this issue, this study made several important technical innovations: first, upgrading the 1D navigator to a 2D navigator to cope with potential severe involuntary motion during human brain scans; second, adopting GRAPPA-based accelerated acquisition to shorten the readout time; and finally, using a diffusion preparation technique instead of directly inserting diffusion gradients in the GRASE readout, allowing the sequence to flexibly adapt to different pulse and vibration gradient requirements. In addition, although GRASE encoding combines the advantages of FSE and EPI encoding, 3D GRASE encoding also inherits the drawbacks of both. Various higher-order echo signals inherent in GRASE encoding can affect diffusion parameters. However, after applying the multi-shot acquisition technique, the number of refocusing pulses in GRASE encoding is greatly reduced, and the higher-order echo signals of GRASE encoding did not cause deviations in FA and ADC, as shown in Fig. 4.10.

Through the comparative analysis of 2D and 3D sequences, this study further evaluated the impact of 2D and 3D acquisition on fiber tracking and microstructure

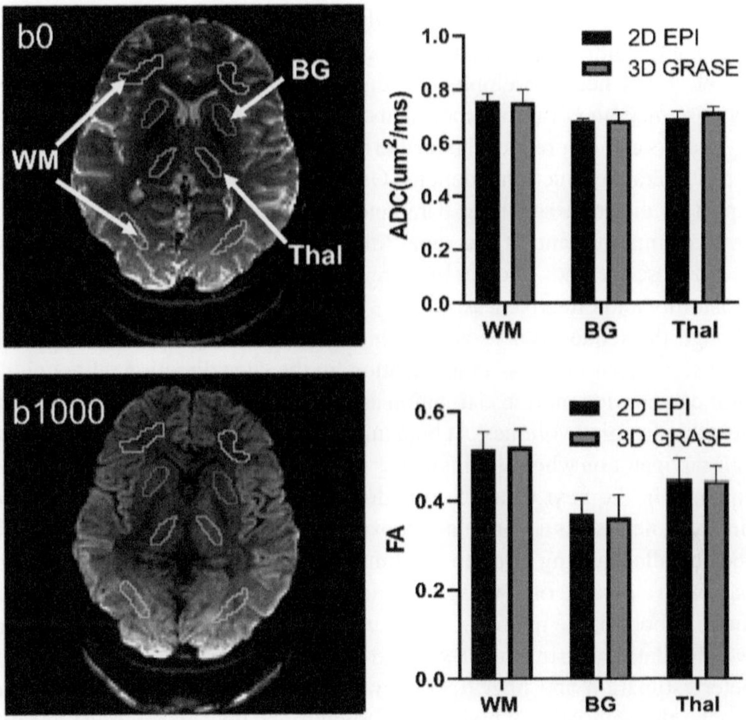

Fig. 4.10 ADC and FA of 3D GRASE and 2D EPI sequences scanning images. The compared ROIs included white matter (WM, blue), basal ganglia (BG, pink), and thalamus (Thal, green). In all ROIs, there was no significant difference in ADC and FA between the 3D GRASE and 2D EPI sequences ($p > 0.1$)

parameter estimation, and found a significant statistical difference in the fiber bundle cross-sectional area in the midbrain of the corticospinal tract [23–25]. Since the corticospinal tract is mainly distributed along the superior-inferior direction (slice direction), the poor slice profile of the 2D EPI sequence may lead to deviations in fiber estimation [16]. Furthermore, this study also found differences in the distribution of fibers in cortical regions between the two sequences. Compared with the 2D EPI sequence, the images scanned by the 3D sequence had higher slice resolution and higher SNR, which could more accurately detect cortical fibers. This finding is consistent with previous studies, which revealed that images acquired using a 3D RARE sequence showed more fibers in the frontal lobe region compared to the 2D EPI sequence [7]. This study found that the 3D sequence was more accurate both in more accurately estimating fiber connectivity and in discovering the number of fibers in cortical regions. Accurate nerve fiber tracking results not only help identify the brain's neural network and better understand the close relationship between brain structure and function, such as studying how the brain internally processes information and how the connections between different brain regions affect behavior and cognitive functions, but are also crucial for the early diagnosis and treatment progress evaluation of many neurodegenerative diseases [26] (Parkinson's disease, Alzheimer's disease, etc.) and neurodevelopmental disorders [27] (autism, etc.). At the same time, precise mapping of nerve fibers can even assist neurosurgical procedures [28], helping doctors avoid or reduce damage to key neural pathways, thereby reducing the risk of postoperative functional deficits. Our future research will focus on exploring the impact of different scanning parameters on fiber tracking and microstructure parameter estimation, and at the same time, continuously optimize 3D sequences and extend these methods to other brain regions and disease models to further verify their feasibility and effectiveness.

Although the sequence proposed in this chapter has been greatly improved and has made significant progress in application, in the slice direction, although the 2D navigator can provide motion estimation and phase correction, it is only suitable for relatively thin imaging volumes. At high image resolution, the phase correction effect of the 2D navigator on whole-brain images is limited. This study will further optimize the sequence in the next Chap. 5 to address this issue. In addition, the choice of scanning parameters has a significant impact on microstructure parameter estimation and fiber bundle tracking. Due to the limitation of scanning time, this study only used data with a b-value of 1000 s/mm^2. However, higher b-values can provide more information about tissue microstructure, and future research will consider acquiring more b-value data and using the FBA process to achieve more accurate microstructure parameter estimation and fiber tracking results. This will help to more accurately interpret the characteristics of brain microstructure and fiber connections, and provide more powerful and reliable tools for clinical diagnosis and neuroscience research.

4.6 Conclusion

The twin-navigator-based multi-shot acquisition 3D OGprep-GRASE and 3D PGprep-GRASE sequences proposed in this chapter have to some extent corrected the phase errors between multi-shot acquisitions. By comparing the SNR of images acquired by the twin-navigator-based multi-shot acquisition 3D GRASE sequence and the 2D EPI sequence under the same scanning time, this study found that the 3D GRASE sequence significantly improved image SNR. More profoundly, by comparing the results of fiber tracking and microstructure parameter estimation between the 3D GRASE and 2D EPI sequences under the same SNR, this study found that the 3D sequence can discover more abundant cortical fibers and a higher fiber bundle cross-sectional area in the corticospinal tract. These findings confirm the important value of the twin-navigator-based multi-shot acquisition 3D GRASE sequence proposed in this study in whole-brain TDDMRI research and its application potential in future neuroscience research.

References

1. Wu W, Poser BA, Douaud G et al (2016) High-resolution diffusion MRI at 7T using a three-dimensional multi-slab acquisition. Neuroimage 143:1–14
2. Holtrop JL, Sutton BP (2016) High spatial resolution diffusion weighted imaging on clinical 3 T MRI scanners using multislab spiral acquisitions. J Med Imag 3(2):023501
3. Bruce IP, Chang H-C, Petty C et al (2017) 3D-MB-MUSE: A robust 3D multi-slab, multi-band and multi-shot reconstruction approach for ultrahigh resolution diffusion MRI. Neuroimage 159:46–56
4. Dai E, Liu S, Guo H (2021) High-resolution whole-brain diffusion MRI at 3T using simultaneous multi-slab (SMSlab) acquisition. Neuroimage 237:118099
5. Liao C, Bilgic B, Tian Q et al (2021) Distortion-free, high-isotropic-resolution diffusion MRI with gSlider BUDA-EPI and multicoil dynamic B0 shimming. Magnet Reson Med 86(2):791–803
6. Setsompop K, Fan Q, Stockmann J et al (2018) High-resolution in vivo diffusion imaging of the human brain with generalized slice dithered enhanced resolution: Simultaneous multislice (g S lider-SMS). Magnet Reson Med 79(1):141–151
7. Frank LR, Jung Y, Inati S et al (2010) High efficiency, low distortion 3D diffusion tensor imaging with variable density spiral fast spin echoes (3D DW VDS RARE). Neuroimage 49(2):1510–1523
8. Jezzard P, Barnett AS, Pierpaoli C (1998) Characterization of and correction for eddy current artifacts in echo planar diffusion imaging. Magnet Reson Med 39(5):801–812
9. Wu D, Liu D, Hsu YC et al (2021) Diffusion-prepared 3D gradient spin-echo sequence for improved oscillating gradient diffusion MRI. Magnet Reson Med 85(1):78–88
10. Prasloski T, Rauscher A, Mackay AL et al (2012) Rapid whole cerebrum myelin water imaging using a 3D GRASE sequence. Neuroimage 63(1):533–539
11. Raffelt DA, Tournier J-D, Smith RE et al (2017) Investigating white matter fibre density and morphology using fixel-based analysis. Neuroimage 144:58–73
12. Tao Z, Yi S, Dan W et al (2021) Accelerated chemical exchange saturation transfer acquisition by joint K-space and image-space parallel imaging (KIPI)
13. Song AW, Chang H-C, Petty C et al (2014) Improved delineation of short cortical association fibers and gray/white matter boundary using whole-brain three-dimensional diffusion tensor imaging at submillimeter spatial resolution. Brain Connect 4(9):636–640
14. Li H, Zu T, Hsu YC et al (2023) Inversion-recovery-prepared oscillating gradient sequence improves diffusion-time dependency measurements in the human brain. J Magnet Reson Imag 57(2):446–453
15. Engström M, Mårtensson M, Avventi E et al (2015) On the signal-to-noise ratio efficiency and slab-banding artifacts in three-dimensional multislab diffusion-weighted echo-planar imaging. Magnet Reson Med 73(2):718–725
16. Engström M, Skare S (2013) Diffusion-weighted 3D multislab echo planar imaging for high signal-to-noise ratio efficiency and isotropic image resolution. Magnet Reson Med 70(6):1507–1514
17. Shao X, Ma SJ, Casey M et al (2019) Mapping water exchange across the blood–brain barrier using 3D diffusion-prepared arterial spin labeled perfusion MRI. Magnet Reson Med 81(5):3065–3079
18. Porter DA, Heidemann RM (2009) High resolution diffusion-weighted imaging using readout-segmented echo-planar imaging, parallel imaging and a two-dimensional navigator-based reacquisition. Magnet Reson Med Off J Int Soc Magnet Reson Med 62(2):468–475
19. Moeller S, Ramanna S, Lenglet C et al (2020) Self-navigation for 3D multishot EPI with data-reference. Magnet Reson Med 84(4):1747–1762
20. Mori S, Van Zijl PC (1998) A motion correction scheme by twin-echo navigation for diffusion-weighted magnetic resonance imaging with multiple RF echo acquisition. Magnet Reson Med 40(4):511–516

21. Aggarwal M, Mori S, Shimogori T et al (2010) Three-dimensional diffusion tensor microimaging for anatomical characterization of the mouse brain. Magnet Reson Med 64(1):249–261
22. Wu D, Xu J, Mcmahon MT et al (2013) In vivo high-resolution diffusion tensor imaging of the mouse brain. Neuroimage 83:18–26
23. Jang SH (2014) The corticospinal tract from the viewpoint of brain rehabilitation. J Rehabil Med 46(3):193–199
24. Lemon RN, Griffiths J (2005) Comparing the function of the corticospinal system in different species: organizational differences for motor specialization? Muscle Nerve Off J Am Assoc Electrodiagnostic Med 32(3):261–279
25. Lemon RN (2008) Descending pathways in motor control. Annu Rev Neurosci 31:195–218
26. Zhang X, Sun Y, Li W et al (2019) Characterization of white matter changes along fibers by automated fiber quantification in the early stages of Alzheimer's disease. NeuroImage Clin 22:101723
27. Wilkinson M, Wang R, Van Der Kouwe A et al (2016) White and gray matter fiber pathways in autism spectrum disorder revealed by ex vivo diffusion MR tractography. Brain Behav 6(7):e00483
28. Fernandez-Miranda JC, Pathak S, Engh J et al (2012) High-definition fiber tractography of the human brain: neuroanatomical validation and neurosurgical applications. Neurosurgery 71(2):430–453
29. Li H, Hsu YC, Zu T et al (2023) 3D diffusion MRI with Twin-navigator-based GRASE for comparison of fiber-tracking using 2D and 3D sequences in human whole brain [J]

Chapter 5
3D Navigator based 3D Gradient Spin Echo Imaging

Abstract This chapter presents the development and application of 3D navigator-based multi-shot acquisition 3D oscillating gradient prepared gradient and spin echo (3D OGprep-GRASE) and pulsed gradient prepared gradient and spin echo (3D PGprep-GRASE) sequences for whole-brain high-resolution time-dependent diffusion MRI (TDDMRI) on a clinical 3 T system. A key innovation is the integration of a 3D navigator to effectively correct inter-segment phase errors inherent in multi-shot acquisitions, outperforming 1D and 2D navigators. The sequences also incorporate 3D GRAPPA acceleration in both phase-encoding directions to enhance acquisition efficiency and image quality. This allowed for the detailed investigation of the diffusion time dependence of ADC in the cerebral cortex gray matter at high resolution. The study utilized a power-law model to characterize this dependence, yielding an exponent θ, which showed regional variations across the cortex. Furthermore, a correlation analysis revealed a negative relationship between θ and microstructural parameters derived from NODDI (ICVF, ODI) and DKI (MK, RK) models, suggesting links between microstructural complexity and diffusion characteristics.

Keywords 3D navigator · Multi-shot acquisition · Whole-brain TDDMRI · Cerebral cortex · Diffusion time dependence · Power-law model · 3D GRAPPA

5.1 Introduction

The content of Chap. 4 clarified that the key factor for achieving whole-brain high-resolution 3D TDDMRI is to efficiently segment the 3D k-space data into multiple segments during multi-shot acquisition, and most importantly, it is essential to accurately correct the inter-segment phase errors caused by subject motion before merging these data. To achieve this goal, various phase correction methods have been previously proposed, among which k-space navigators have demonstrated excellent correction effects when dealing with multi-shot acquisition data [1–4]. However, current research has focused on 1D navigators that only acquire signals along the readout direction or 2D navigators that only acquire phase encoding

D. Wu et al., *Development and Application of 3D Diffusion Magnetic Resonance Imaging Sequences*, Advanced Topics in Science and Technology in China 5,
https://doi.org/10.1007/978-981-95-0808-2_5

in a single direction [5, 6], resulting in limited correction effectiveness. For 3D TDDMRI sequences, exploring the application of 3D navigators in whole-brain imaging will be an extremely difficult challenge. Building upon the work of the twin navigator-based multi-shot acquisition 3D OGprep-GRASE and PGprep-GRASE sequences developed in Chap. 4, this study further developed multi-shot acquisition 3D OGprep-GRASE and PGprep-GRASE sequences integrated with a 3D navigator. It can stably acquire TDDMRI images on a clinical 3 T MRI system, and simultaneously, combined with 3D GRAPPA acceleration technology accelerated in two phase encoding directions, significantly improving the acquisition efficiency of the sequence.

Accurately quantifying the diffusion time dependence of ADC in the gray matter of the cerebral cortex has always been an important and challenging research task. Although there have been previous related discussions, due to the complexity of the cerebral cortex gray matter structure and the image resolution constraints faced in these structures, the results have not been quantitatively evaluated. However, with the newly developed technology in this study: 3D multi-shot acquisition OGprep-GRASE and PGprep-GRASE sequences based on a 3D navigator, it is possible to deeply quantify the diffusion time dependence of cortical gray matter ADC. Mitra et al. [7] and Parsons et al. [8] summarized the power-law model of the diffusion coefficient's time dependence in TDdMRI: as diffusion time increases, under the tortuosity limit, weakly heterogeneous media exhibit local fluctuations, and their diffusion time dependence has a perturbation solution. Novicov et al. [9] demonstrated that the diffusion coefficient changing with diffusion time can distinguish the universality classes of mesoscopic structural complexity. The significance of θ lies in helping researchers determine the dimension of diffusion and the type of disorder. After determining the type of disorder, the most reasonable microstructure model can be found in that class. This study also explored the association between θ based on the power-law model [9, 10] and some advanced dMRI model parameters. For example, microstructural indices obtained from neurite orientation dispersion and density imaging [11] (NODDI) and DKI [12].

5.2 Theory

5.2.1 Three-Dimensional Navigator Acquisition Module

As shown in Fig. 5.1, the 3D navigator-based multi-shot acquisition 3D OGprep-GRASE sequence proposed in this chapter mainly consists of five modules, including a global stabilization module, a diffusion preparation module, a fat saturation module, a 3D GRASE acquisition module, and a 3D navigator acquisition module. At the same time, this chapter also designed a 3D navigator-based multi-shot acquisition PGprep-GRASE sequence.

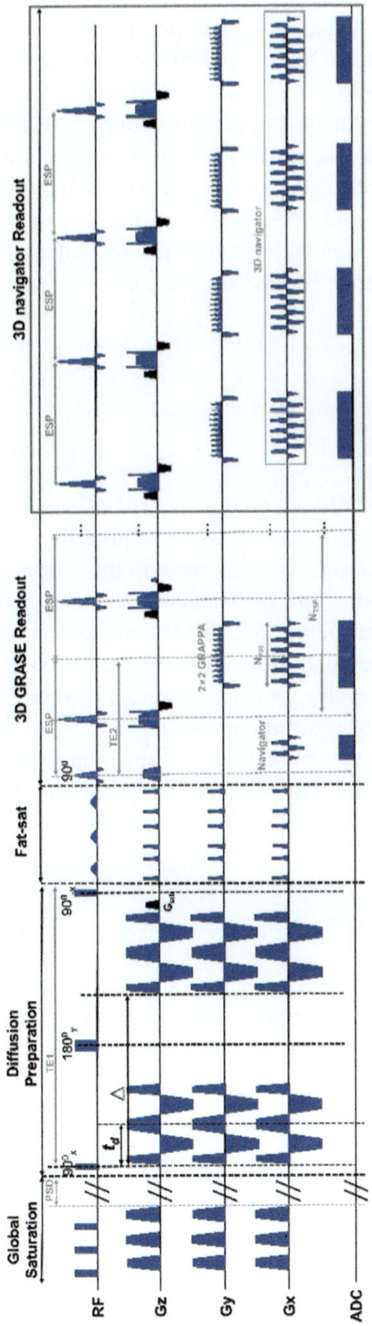

Fig. 5.1 Schematic diagram of the 3D navigator-based multi-shot acquisition 3D OGprep-GRASE sequence framework [13]

In this study, GRAPPA acceleration was performed in both the EPI and FSE phase encoding directions of the 3D GRASE. Simultaneously, the FSE encoding was divided into multi-shot acquisition in an interleaved manner. During the entire acquisition process, the 3D navigator readout module played a crucial role. It is embedded at the end of the readout sequence and is designed as four linearly centered planes in the FSE direction, with phase encoding along the EPI direction (N_{EPI} lines). This design can acquire 3D k-space to generate low-resolution navigator images, which are subsequently used to correct the phase errors of the multi-shot acquired images. To verify the effectiveness of this method, this study comparatively analyzed its difference in phase correction with double 1D and double 2D navigators, and the correction methods for double 1D and double 2D navigators are consistent with Chap. 4 of this study.

5.2.2 *Phase Correction*

For the 3D navigator, this study adopted a method of simultaneously performing phase encoding along the y-axis and z-axis to acquire the navigator readout, and performed phase error correction for each plane in the image domain x–y–z space. First, the k-space data of each acquired segment was reconstructed into an image using the multiplexed sensitivity encoding (MUSE) technique from Chap. 3 of this study, and the navigator data corresponding to each segment of k-space data was reconstructed into a navigator image by zero-filling around it. Then, the navigator images were used to correct the phase difference of the corresponding images between different segments. Let $\varphi_{n,s}(x, y, z)$ be the navigator image corresponding to the k-space signal obtained from the s-th pulse excitation, and $\psi_{i,s}(x, y, z)$ be the image corresponding to the k-space signal obtained from the s-th pulse excitation. The formula for calculating the phase-corrected $\psi'_{i,s}$ (x, y, z) is as follows:

$$\psi'_{i,s}(x, y, z) = \psi_{i,s}(x, y, z) - \varphi_{n,s}(x, y, z) \tag{5.1}$$

Finally, $\psi'_{i,s}$ (x, y, z) is transformed back to k-space, and the signal at the k-space position corresponding to the s-th pulse excitation is extracted. This completes the phase correction of the s-th pulse excitation.

5.2.3 *3D GRAPPA Accelerated Imaging*

This study has already introduced the use of 2D GRAPPA acceleration with the 3D GRASE sequence in Chap. 4. In this chapter, the sequence was further accelerated by simultaneously performing acceleration in both the EPI and FSE directions. Figure 5.2 shows the k-space trajectory of the sequence with 3D GRAPPA 2×2 acceleration and 3D navigator acquisition, taking a 4-shot acquisition as an example.

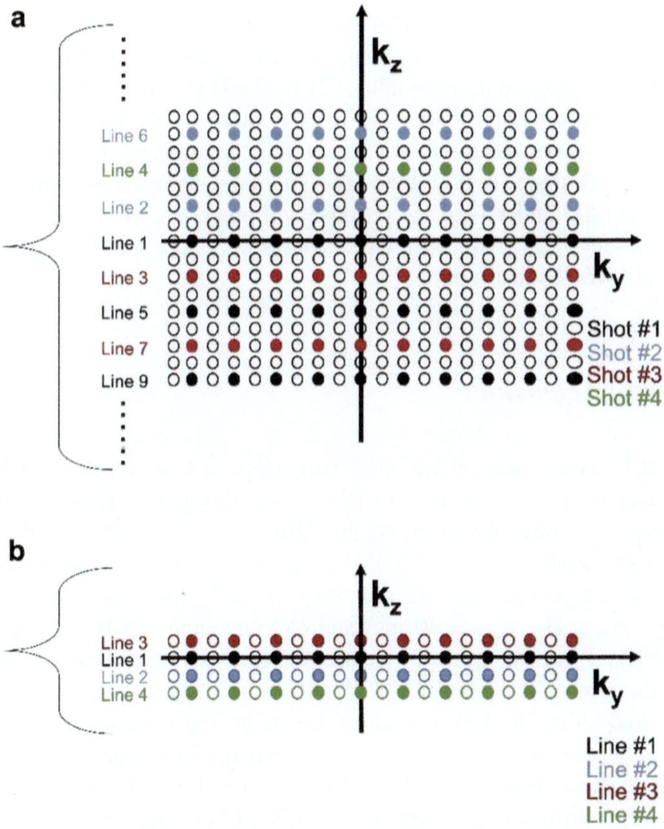

Fig. 5.2 Schematic diagram of the k-space sampling trajectory of the 3D GRASE sequence in the k_z–k_y space. **a** The image is fully sampled on k_x, and k_y and k_z are subjected to 2×2 GRAPPA acceleration. Solid and hollow circles represent the measured and missing k-space data in each segment, respectively. In the centrally encoded k_z direction, $k_z = 0$ is the starting line (Line 1), and the adjacent lines (Line 2, 3, 4) are the first readout of the corresponding segment. Line $1 + 4n$ to Line $4 + 4n$ ($n = 0, 1, 2, 3$) are respectively assigned to Shot 1–4. **b** The trajectory diagram of the 3D navigator in the k_z and k_y spaces. Only the central four planes in the k_z direction are acquired

5.3 Experimental Methods

5.3.1 Magnetic Resonance Experiments

This study used in-house developed 3D navigator-based multi-shot acquisition 3D OGprep-GRASE and 3D PGprep-GRASE sequences for data acquisition. A total of seven healthy volunteers were recruited (average age = 22 ± 2 years, range 20–25 years, all male). All scans were performed on a Siemens 3 T MAGNETOM Prisma MRI scanner equipped with a 64-channel head coil. The maximum gradient strength of this scanner is 80 mT/m, and the maximum slew rate is 200 mT/m/ms. This study set three important goals: (1) To evaluate the impact of adding the proposed 3D navigator on image quality. (2) To assess the diffusion time dependence of the cerebral cortex gray matter. (3) To fit θ based on the power-law model and explore its correlation with microstructural feature parameters. Based on these three goals, this study designed two targeted experimental schemes. Before the experiment, all volunteers fully understood the purpose and procedure of the study and signed informed consent forms. All research procedures were approved by the ethics review board of the affiliated institution.

5.3.2 Data Acquisition

Experiment 1: Assessment of Diffusion Time Dependence of Cerebral Cortex Gray Matter: 3D navigator-based multi-shot acquisition 3D OGprep-GRASE and PGprep-GRASE sequences were used to assess the diffusion time dependence of ADC within the cortical gray matter regions. TDDMRI signals were acquired using PG encoding ($\delta/\Delta = 15/25$ ms, 0 Hz, $\Delta_{\text{eff}} = 20$ ms; $\delta/\Delta = 15/35$ ms, 0 Hz, $\Delta_{\text{eff}} = 30$ ms; $\delta/\Delta = 15/45$ ms, 0 Hz, $\Delta_{\text{eff}} = 40$ ms) and OG encoding (25 Hz, 2 cycles, $\Delta_{\text{eff}} \approx 10$ ms; 50 Hz, 4 cycles, $\Delta_{\text{eff}} \approx 5$ ms) frequencies. All other scanning parameters remained the same: TR/TE1/TE2 = 2600/84/46 ms, $N_{\text{EPI}} = 67$, $N_{\text{FSE}} = 4$, with 12 shots acquired along the FSE encoding direction. The readout matrix size of the navigator echo was the same as the imaging matrix in the x and y directions, with NFSE = 4 in the Z direction. FOV = $220 \times 200 \times 144$ mm^3, resolution = $1.5 \times 1.5 \times 1.5$ mm^3, diffusion directions = 6, number of averages = 2, $b = 500$ s/mm^2, $b = 0$ averaged 4 times, GRAPPA acceleration factor set to 2×2. The scan time for each diffusion time was 5 min. In this study, 3D T1-weighted images of the brain were acquired for cortical gray matter segmentation. Since the multi-shot acquisition 3D PGprep-GRASE sequence is more susceptible to aliasing artifacts, one volunteer was also selected to undergo PG encoding acquisition ($\delta/\Delta = 15/35$ ms, 0 Hz, $\Delta_{\text{eff}} = 30$ ms), TE = 76 ms, $b = 1000$ s/mm^2, number of diffusion directions = 6, using 1D, 2D, and 3D navigator-based multi-shot acquisition 3D PGprep-GRASE sequences respectively, to compare the phase correction effects of double 1D, double 2D, and 3D navigators. Other parameters remained unchanged. In addition, one volunteer was also selected to undergo dMRI signal acquisition using a 2D EPI sequence with OG encoding (50 Hz frequency), resolution = $2 \times 2 \times 2$ mm^3, number of averages = 4, TR = 25 s, and scan time = 12 min, to compare the improvement in image resolution and scan speed of 3D GRASE compared to 2D EPI. Other parameters remained unchanged.

Experiment 2: Correlation between θ and microstructural feature parameters: This experiment aimed to correlate θ with several microstructural features fitted by the NODDI and DKI models. The experimental data were selected from the HCP dataset, including 10 healthy subjects (average age = 22 ± 2 years, range 20–25 years, all male). The intra cellular volume fraction (ICVF) and orientation dispersion index

(ODI) were extracted from the NODDI model, and the axial kurtosis (AK), mean kurtosis (MK), and radial kurtosis (RK) were extracted from the DKI model. The data were acquired using PG encoding, with the following parameters: FOV = 181 × 217 × 181 mm^3, resolution = 1.25 × 1.25 × 1.25 mm^3, $b = 1000$, 2000, and 3000 s/mm^2, number of diffusion directions = 90, number of averages = 1, and $b = 0$ images were averaged 18 times.

5.3.3 *Data Analysis*

K-space data reconstruction: The raw *k*-space data acquired from the 3D GRASE sequence underwent a reconstruction process based on 3D GRAPPA, supplemented by 3D navigator phase correction. The reconstructed images went through several preprocessing steps, including Gibbs artifact removal, eddy current correction, image registration, and bias field correction. All methods were performed using the MRtrix software package.

Calculation of ADC maps and segmentation of cerebral cortical gray matter: First, the ADC of the reconstructed images was calculated. Then, for the segmentation of the cerebral cortical gray matter, this study registered T1-weighted images to the Automated Anatomical Labeling (AAL) [14] atlas, followed by precise registration of the aligned images to the dMRI data. The ADC values at multiple diffusion times within the cortical gray matter were fitted to a power-law model to estimate θ. According to the theoretical framework, a higher θ indicates lower structural disorder. As shown in Eq. 5.2:

$$D_t = c \times t_d^{\theta} + D_{t0} \tag{5.2}$$

Microstructural analysis: Experiment 2 used the NODDI Matlab toolbox to calculate the NODDI indices (ICVF and ODI) for the 10 healthy subjects. The DKI indices (AK, MK, RK) were obtained in the cerebral cortical gray matter regions using weighted linear least squares estimation [8]. Subsequently, comprehensive statistical analyses were performed using Pearson correlation coefficients in MATLAB to evaluate the correlation between the obtained NODDI and DKI indices and the θ values. The purpose of this analysis method was to establish meaningful links between diffusion exponents and microstructural features.

5.4 Experimental Results

Figure 5.3 shows the DWI images acquired using the PGprep-GRASE sequence with 3D multi-shot acquisition based on a 3D navigator. The images display the changes before and after phase correction using twin 1D, twin 2D, and 3D navigators at a *b*-value of 1000 s/mm^2 from three different orientations: axial, coronal, and

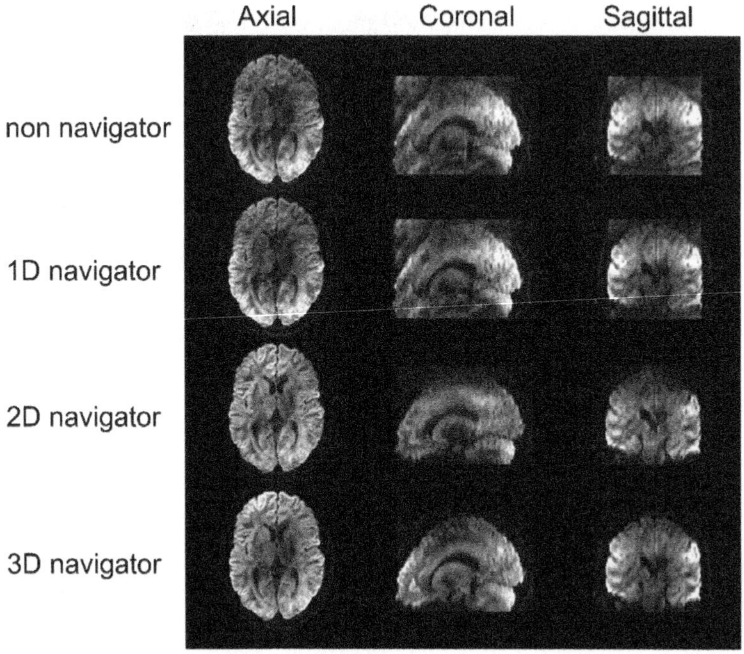

Fig. 5.3 Images before and after phase correction with twin 1D, twin 2D, and 3D navigators. Three-dimensional view data obtained at a *b*-value of 1000 s/mm^2 in one diffusion direction

sagittal. This study found that directly reconstructed DWI was significantly affected by phase errors. These errors caused image blurring and affected the image quality and accuracy of diffusion parameters. However, navigator-based phase correction greatly improved this situation, contributing to enhanced image quality and accurate structural depiction. Notably, the 3D navigator showed a more significant advantage in reducing residual artifacts compared to twin 1D and twin 2D navigators, especially in the FSE phase-encoding direction. This sequence provides important technical support for more precise lesion detection and microstructural parameter evaluation in clinical applications.

Figure 5.4 shows the DWI images acquired using the OGprep-GRASE sequence with 3D multi-shot acquisition based on a 3D navigator and the 2D OG-EPI sequence at a *b*-value of 500 s/mm^2. Comparison of the SNR of the images in the white matter (WM, blue), basal ganglia (BG, pink), and thalamus (Thal, green) regions of interest (ROI) shows that the 3D sequence achieved the same SNR as the 2D sequence while improving the image resolution by 1.37 times and the scan time by 1.4 times.

This study used the 3D OGprep-GRASE and PGprep-GRASE sequences with 3D multi-shot acquisition based on a 3D navigator to quantify the ADC of different gray matter regions of the cerebral cortex at a 1.5 mm isotropic resolution. The ROIs included the parietal lobe, occipital lobe, precentral gyrus, postcentral gyrus, frontal lobe, temporal lobe, insula, cingulate gyrus, and precuneus. The diffusion

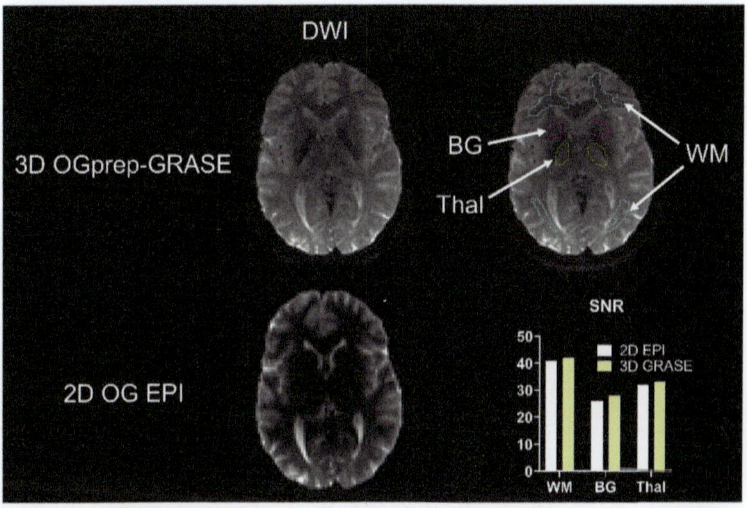

Fig. 5.4 Image SNR of 3D OGprep-GRASE and 2D OG-EPI sequences. The ROIs for image comparison at a *b*-value of 500 s/mm^2 include white matter (WM, blue), basal ganglia (BG, pink), and thalamus (Thal, green)

time-dependency of these regions was evaluated by fitting a power-law model to the ADC values at different diffusion times, as shown in Fig. 5.5a. In all ROIs, θ exceeded 0.5. The highest θ value was found in the occipital lobe, followed by the precentral and postcentral gyri, while the lowest θ values were observed in the temporal lobe, insula, and frontal lobe. Figure 5.5b shows the mapping of θ values in the cortical gray matter regions onto a cortical gray matter surface template. This mapping further indicated higher θ values in sensory cortices (such as the parietal and occipital lobes) and lower θ values in higher-order cortices (such as the frontal and temporal lobes). This result suggests a higher degree of microstructural disorder in the higher-order cortical regions of the brain, providing a new perspective for a deeper understanding of cortical structure and functional differences between different cortical regions.

This study performed a correlation analysis between θ and various microstructural parameters fitted by the NODDI and DKI models. Through Pearson correlation analysis, this study found a significant negative correlation ($p < 0.05$) between θ and the NODDI model parameters intra cellular volume fraction (ICVF) and orientation dispersion index (ODI), as shown in Fig. 5.6a. This indicates a relationship between the decrease in θ values and the increase in neuronal cell density and orientation dispersion, providing a new approach to understanding the complexity of neuronal microstructure. Furthermore, a negative correlation was also found between θ and the DKI model parameters mean kurtosis (MK) and radial kurtosis (RK), as shown in Fig. 5.6b. This implies a significant correlation between highly non-Gaussian diffusion and microstructural disorder. Notably, this study found the highest Pearson correlation coefficient between θ and ODI, suggesting a strong relationship between the degree of neuronal orientation dispersion and structural disorder.

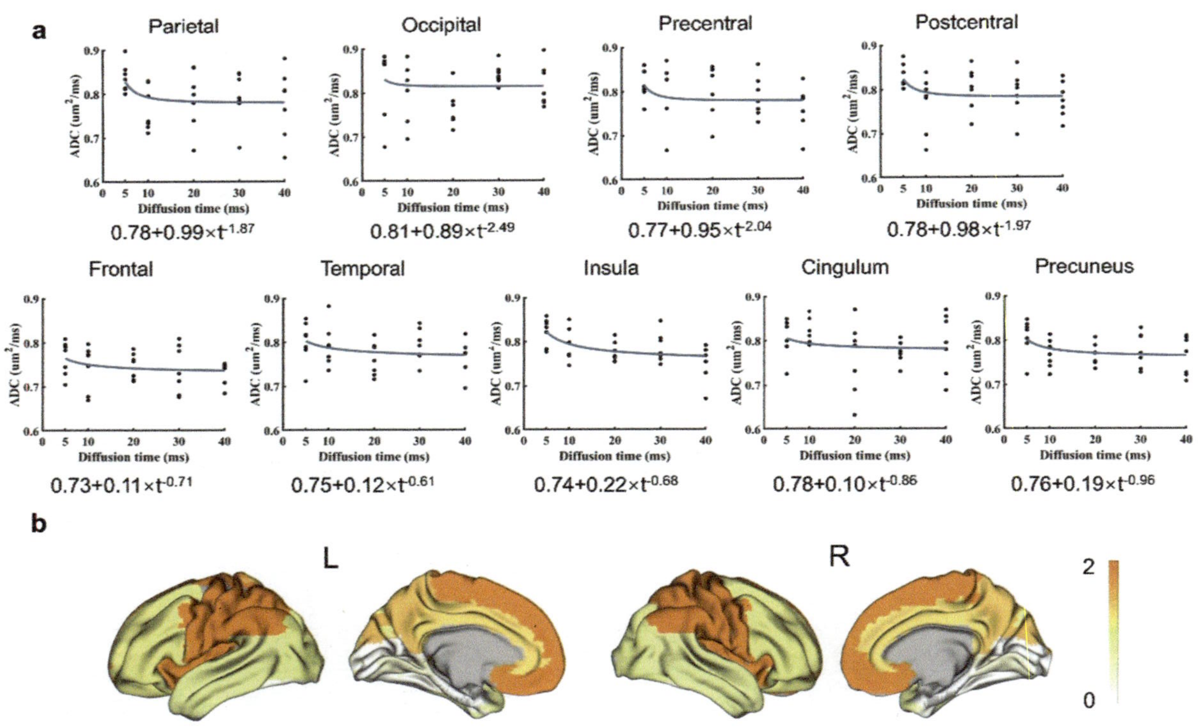

Fig. 5.5 The fitting results based on the power-law model of the cerebral cortex at different diffusion times. **a** ADC values measured at different diffusion times (5, 10, 20, 30, 40 ms) using the 3D GRASE sequence with a 3D navigator. **b** Mapping of θ values in the cortical gray matter regions based on the power-law model fitting and projected onto a cortical gray matter surface template. Data from seven volunteers were used for fitting

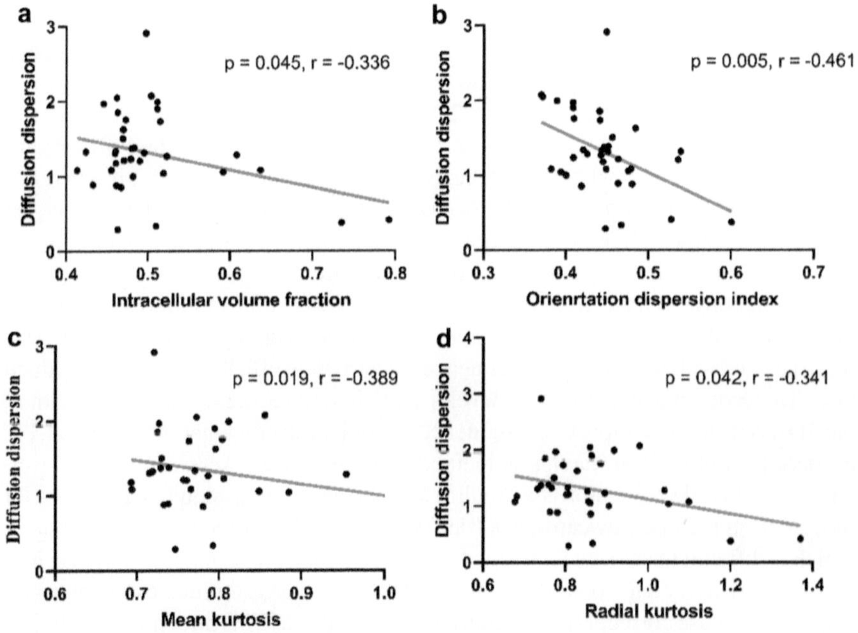

Fig. 5.6 Pearson correlation between θ and NODDI and DKI model microstructural parameters. θ is negatively correlated ($p < 0.05$) with **a** microstructural parameters fitted by the NODDI model (ICVF, ODI) and **b** microstructural parameters fitted by the DKI model (MK, RK)

5.5 Discussion

This study developed 3D OGprep-GRASE and 3D PGprep-GRASE sequences with 3D multi-shot acquisition based on a 3D navigator, specifically designed for high-resolution TDDMRI scanning on clinical 3 T MRI scanners, significantly enhancing the ability to characterize diffusion time-dependency in the cortical gray matter regions. A key factor in achieving this goal was the effective phase correction of errors between each RF pulse excitation by the 3D navigator. During acquisition, the use of 3D GRAPPA acceleration technology, simultaneously accelerating in both phase-encoding directions, was crucial not only for reducing TE and TR but also for improving data acquisition efficiency and image quality. Furthermore, this study found a negative correlation between θ and ICVF, ODI, MK, and RK through quantitative correlation analysis. This contributes to understanding the relationship between neuronal microstructure and neurodegenerative diseases, providing valuable references and methods for future research and clinical diagnosis.

In related research and clinical practice, 3D dMRI sequences have demonstrated advantages in improving image SNR and layer profiles [15–17]. For example, the 3D OGprep-GRASE sequence designed in Chap. 3 can improve image SNR and shorten scan time on clinical scanners. However, limitations such as limited image resolution, long scan times, and inconsistent phases between RF pulses pose significant challenges for whole-brain acquisition with these methods. Chapter 4 proposed a 3D OGprep-GRASE sequence with twin navigator-based multi-shot acquisition for whole-brain TDdMRI. The twin 2D navigator in that sequence could correct phase errors between RF pulses. However, its correction effect in the FSE direction was limited, especially when the image *b*-value and resolution were high. Therefore, developing a 3D navigator suitable for 3D TDDMRI sequences has become urgent, as it can not only overcome the aforementioned challenges but also help to deeply understand the brain's organizational microstructure.

The challenge of using a 3D navigator is the excessive time required for its acquisition compared to the data of each segmented *k*-space. Although more accurate phase information can be obtained, the sequence's acquisition efficiency is greatly reduced [18]. Therefore, it is imperative to develop a method to reduce the acquisition time of the 3D navigator and improve the sequence's acquisition efficiency. McNab et al. [19] proposed a multi-TR 3D navigator method to correct steady-state DWI phase errors. However, this 3D navigator is a "multi-shot navigator" that requires assuming that most motion is caused by cardiac pulsation and that this motion is highly repeatable, which is difficult to achieve in practice. This study added a 3D navigator at the end of the 3D GRASE module, obtaining low-resolution 3D navigator images for multi-shot acquisition phase correction by acquiring the central four FSE direction encodings in *k*-space. Although only four phase-encoded planes were acquired, experiments have shown that the phase correction effect of the 3D navigator is significantly better than that of the 2D navigator.

The power-law model of the diffusion coefficient at short diffusion times can help researchers determine the dimension and type of disorder of diffusion. After determining the type of disorder, the most reasonable microstructural model in that category can be found for parameter fitting, thereby obtaining important information about tissue microstructure. For example, $\theta = 1/2$ has been found in both normal and large-area ischemic rat brains, consistent with water diffusion along narrow neurons. This suggests that increased disorder after the onset of global cerebral ischemia in rats may originate from beading of dendrites and axons caused by ischemia [9, 20]. The power-law model mainly provides theoretical support for water diffusion patterns in the brain and tumors. $\theta = 1/2$ has been confirmed in healthy and globally ischemic rodent brain tissue [9, 21, 22]. Recently, a study on prostate cancer found $\theta = 1$ fitted by the power-law model in the tumor peripheral region [23], while another study found $\theta = 1/2$ fitted by the power-law model in epithelial cells and fibromuscular stroma [24]. This may reveal the different microstructural characteristics between prostate regions, aiding in the early diagnosis of prostate cancer. Using the 3D GRASE sequence with 3D multi-shot acquisition based on a 3D navigator, this study conducted a detailed discussion on the correlation between θ and microstructural parameters fitted by the NODDI and DKI models. Consistent with recent studies [20, 25], this study observed significant diffusion time-dependency in different regions of the brain. The power-law model calculated θ values >0.5 in most gray matter regions, indicating more ordered microstructural organization in these areas. Notably, lower θ values were measured in higher-order cortices, reflecting a higher degree of structural disorder compared to primary cortices. This finding was further validated by the negative correlation between θ and microstructural parameters fitted by the NODDI and DKI models, revealing that complex microstructural organization is associated with higher neuronal density, orientation dispersion, and non-Gaussian diffusion. This helps us further deepen our understanding and exploration of the brain's organizational microstructure. Therefore, the high-resolution TDDMRI technique proposed in this study provides a new means for studying and depicting the cortical microenvironment. These findings are of significant value both theoretically and practically, providing new research directions for the further development of neuroimaging and clinical practice. These research results lay a solid foundation for future in-depth studies on the interrelationship between neuronal microstructure and brain diseases, having a profound impact on promoting the realization of precision medicine and personalized treatment.

Despite the significant breakthroughs achieved in this study, there are still two current limitations. First, limited by the scan time, Experiment 2 could only use the results of 5 diffusion time points to estimate θ. Future studies will include more diffusion times, which will help to more comprehensively interpret the time-dependency of diffusion parameters. Second, the TDDMRI image resolution of this study was 1.5 mm isotropic. To obtain accurate cortical gray matter segmentation, TDDMRI images need to be registered with 1 mm isotropic T_1-weighted images. If the resolution of TDDMRI can be further improved, it will provide more accurate results for the estimation of θ. Despite the above two limitations, the findings of this study have undoubtedly made significant contributions to the neuroimaging field, providing new ideas and possibilities for future TDDMRI research. In the future, with the deepening of research work and technological advancements, the microstructure of the cerebral cortex will be depicted more accurately. Finally, although image scan time and SNR have been improved, the currently clinically achievable oscillating frequency is still limited to ~50 Hz and the b-value is 500 s/mm^2. Advances in magnetic resonance hardware, such as the Connectome scanner [26] and the production of head-only gradient coils, will reduce peripheral nerve stimulation [27], greatly promoting the application of TDDMRI in clinical research.

5.6 Conclusion

The 3D OGprep-GRASE and PGprep-GRASE sequences with 3D multi-shot acquisition based on a 3D navigator proposed in this chapter were applied in a 3 T clinical system for whole-brain high-resolution diffusion time-dependency research. Among these, the 3D navigator corrected the phase errors between multi-shot acquisitions. The diffusion time-dependency based on the power-law model showed obvious regional characteristics, revealing the differences in tissue microstructure across different brain regions. In addition, this study found that complex microstructure often corresponds to higher neurite density, higher orientation dispersion, and non-Gaussian diffusion. These in-depth and detailed studies provide a new perspective for revealing the complex structure of cortical gray matter in the human brain, and also provide new insights and tools for researchers in the field of TDdMRI.

References

1. Porter DA, Heidemann RM (2009) High resolution diffusion-weighted imaging using readout-segmented echo-planar imaging, parallel imaging and a two-dimensional navigator-based reacquisition. Magnet Reson Med Off J Int Soc Magnet Reson Med 62(2):468–475
2. Butts K, De Crespigny A, Pauly JM et al (1996) Diffusion-weighted interleaved echo-planar imaging with a pair of orthogonal navigator echoes. Magnet Reson Med 35(5):763–770
3. Dai E, Liu S, Guo H (2021) High-resolution whole-brain diffusion MRI at 3T using simultaneous multi-slab (SMSlab) acquisition. Neuroimage 237:118099
4. Ohalloran R, Aksoy M, Van A et al (2013) 3D isotropic high-resolution diffusion-weighted MRI of the whole brain with a motion-corrected steady-state free precession sequence. Magnet Reson Med 70(2):466–478
5. Aggarwal M, Mori S, Shimogori T et al (2010) Three-dimensional diffusion tensor microimaging for anatomical characterization of the mouse brain. Magnet Reson Med 64(1):249–261
6. Wu D, Xu J, Mcmahon MT et al (2013) In vivo high-resolution diffusion tensor imaging of the mouse brain. Neuroimage 83:18–26
7. Mitra PP, Sen PN, Schwartz LM (1993) Short-time behavior of the diffusion coefficient as a geometrical probe of porous media. Phys Rev B Condens Matter 47(14):8565–8574
8. Parsons EC, Does MD, Gore JC (2006) Temporal diffusion spectroscopy: theory and implementation in restricted systems using oscillating gradients. Magnet Reson Med 55(1):75–84
9. Novikov DS, Jensen JH, Helpern JA et al (2014) Revealing mesoscopic structural universality with diffusion. Proceed Natl Acad Sci 111(14):5088–5093
10. Arbabi A, Kai J, Khan AR et al (2020) Diffusion dispersion imaging: mapping oscillating gradient spin-echo frequency dependence in the human brain. Magnet Reson Med 83(6):2197–2208
11. Zhang H, Schneider T, Wheeler-Kingshott CA et al (2012) NODDI: practical in vivo neurite orientation dispersion and density imaging of the human brain. Neuroimage 61(4):1000–1016
12. Steven AJ, Zhuo J, Melhem ER (2014) Diffusion kurtosis imaging: an emerging technique for evaluating the microstructural environment of the brain. Am J Roentgenol 202(1):W26–W33
13. Li H, Zhu Q, Lu J et al (2025) Diffusion-weighted GRASE sequence with 3D navigator for high-resolution time-dependent diffusion MRI in the human cortical gray matter [J]. Magnet Reson Med
14. Tzourio-Mazoyer N, Landeau B, Papathanassiou D et al (2002) Automated anatomical labeling of activations in SPM using a macroscopic anatomical parcellation of the MNI MRI single-subject brain. Neuroimage 15(1):273–289
15. Engström M, Mårtensson M, Avventi E et al (2015) On the signal-to-noise ratio efficiency and slab-banding artifacts in three-dimensional multislab diffusion-weighted echo-planar imaging. Magnet Reson Med 73(2):718–725
16. Frank LR, Jung Y, Inati S et al (2010) High efficiency, low distortion 3D diffusion tensor imaging with variable density spiral fast spin echoes (3D DW VDS RARE). Neuroimage 49(2):1510–1523
17. Engström M, Skare S (2013) Diffusion-weighted 3D multislab echo planar imaging for high signal-to-noise ratio efficiency and isotropic image resolution. Magnet Reson Med 70(6):1507–1514
18. Wu W, Miller KL (2017) Image formation in diffusion MRI: a review of recent technical developments. J Magnet Reson Imag 46(3):646–662

19. Mcnab JA, Gallichan D, Miller KL (2010) 3D steady-state diffusion-weighted imaging with trajectory using radially batched internal navigator echoes (TURBINE). Magnet Reson Med Off J Int Soc Magnet Reson Med 63(1):235–242
20. Lee H-H, Papaioannou A, Novikov DS et al (2020) In vivo observation and biophysical interpretation of time-dependent diffusion in human cortical gray matter. Neuroimage 222:117054
21. Does MD, Parsons EC, Gore JC (2003) Oscillating gradient measurements of water diffusion in normal and globally ischemic rat brain. Magnet Reson Med 49(2):206–215
22. Wu D, Zhang J (2017) The effect of microcirculatory flow on oscillating gradient diffusion MRI and diffusion encoding with dual-frequency orthogonal gradients (DEFOG). Magnet Reson Med 77(4):1583–1592
23. Lemberskiy G, Rosenkrantz AB, Veraart J et al (2017) Time-dependent diffusion in prostate cancer. Invest Radiol 52(7):598
24. Lemberskiy G, Fieremans E, Veraart J et al (2018) Characterization of prostate microstructure using water diffusion and NMR relaxation. Front Phys 6:1154
25. Olesen JL, Østergaard L, Shemesh N et al (2022) Diffusion time dependence, power-law scaling, and exchange in gray matter. Neuroimage 251:118976
26. Van Essen DC, Ugurbil K, Auerbach E et al (2012) The human connectome project: a data acquisition perspective. Neuroimage 62(4):2222–2231
27. Tan ET, Shih RY, Mitra J et al (2020) Oscillating diffusion-encoding with a high gradient-amplitude and high slew-rate head-only gradient for human brain imaging. Magnet Reson Med 84(2):950–965

Chapter 6
IR-based 3D Gradient and Spin Echo Imaging

Abstract This study addresses the issue of cerebrospinal fluid (CSF) signal interference in time-dependent diffusion magnetic resonance imaging (TDDMRI), which leads to deviations in apparent diffusion coefficient (ADC) measurements. A novel 3D inversion recovery (IR)-enhanced oscillating gradient-prepared gradient and spin echo (IR-OGprep-GRASE) sequence was proposed to suppress CSF signals by incorporating an IR module prior to the diffusion preparation. This modification significantly improved the accuracy of ADC measurements in brain regions adjacent to ventricles and sulci. Experiments were conducted using 3D GRASE and 2D echo-planar imaging (EPI) sequences on healthy volunteers and glioblastoma patients to evaluate the impact of the IR module on diffusion time-dependent ADC. Results demonstrated that the IR module markedly enhanced ADC diffusion time dependence in hippocampal subfields and tumor regions ($p < 0.01$) and refined the estimation of microstructural parameters such as intracellular fraction and cell diameter. Furthermore, the IR technique exhibited consistent efficacy across different sequences, validating its potential for clinical TDDMRI applications. Despite increased scan duration, this study provides a novel approach to mitigate CSF interference, enhances the reliability of microstructural parameter quantification, and offers critical insights for brain tumor diagnosis and functional neuroimaging research.

Keywords Inversion recovery · 3D GRASE · Time-dependent diffusion MRI · Apparent diffusion coefficient · CSF suppression · Brain tumor microstructure · Hippocampal subfield

6.1 Introduction

As discussed in previous chapters, TDDMRI has become an effective means of detecting and reflecting restricted water diffusion in cellular and axonal tissue structures, capable of characterizing fine microstructures [1]. The OG technique has become an integral part of TDDMRI by virtue of its ability to image over short diffusion times and infer tissue microstructures over a wide range of diffusion times

D. Wu et al., *Development and Application of 3D Diffusion Magnetic Resonance Imaging Sequences*, Advanced Topics in Science and Technology in China 5,
https://doi.org/10.1007/978-981-95-0808-2_6

[2]. However, when targeting human applications, the technique faces many challenges due to the limitation of gradient strengths in practical clinical MRI systems, such as limited oscillation frequencies and *b*-values [3, 4], longer TEs and high duty cycles [5], among others. These problems make the imaging resolution, SNR and contrast of OG technique low. In addition, the analysis of TDDMRI in brain imaging may be compromised due to the high signal intensity and strong diffusion properties of CSF.

As discussed in Sect. 1.4, accurate measurement of ADC is essential for TDDMRI-related studies and microstructural parameter estimation [6, 7]. However, some studies have already found that CSF-triggered PSF and PVE in the sulci and ventricles of the brain can lead to excessive ADC values [8]. This CSF-triggered effect also alters the sensitivity of dMRI for the detection of brain lesions, especially at low *b*-values on DWI For example, Ni et al. [9] reported that although no significant CSF effect was found in ischemic lesions, it significantly altered the ADC in the contralateral brain region, thereby reducing the detection sensitivity. To suppress the effect of CSF in dMRI studies, researchers proposed the IR technique [10]. The 3D OGprep-GRASE sequence proposed in the previous chapters of this study [11], although particularly suitable for TDDMRI imaging on clinical MRI systems, may lead to deviations of the measured ADC from the true value due to the relatively low imaged *b*-value in brain tissue regions adjacent to the ventricles and sulci [8].

To address the problem of inaccurate ADC measurements occurring in regions of brain tissue close to the ventricles and sulci, the present study proposes a conception that the application of an IR module can effectively remove the CSF signal and thus improve the diffusion time-dependent characteristics of these regions. In this study, an IR module was added to the 3D OGprep-GRASE sequence to validate the conception, and the effect of CSF was also evaluated using 3D and 2D sequences, respectively, to reveal whether the problem exists in all types of sequences and whether the IR module is consistently effective in all types of sequences. In addition, this study explored the application of sequences in brain tumors to explore the potential value of the IR module for TDDMRI in the clinic.

6.2 Theory

6.2.1 IR Module

The 3D OGprep-GRASE sequence proposed in the previous chapters of this study has a global stabilization module before the diffusion preparation module, fat saturation module and GRASE readout module. Based on it, the 3D IR-OGprep-GRASE sequence is proposed in this chapter, which uses an IR module to replace the previous global stabilization module, and the subsequent diffusion preparation module, fat saturation module, and GRASE readout module are kept the same as shown in Fig. 6.1. The role of the IR module is to eliminate the CSF signals through the use of

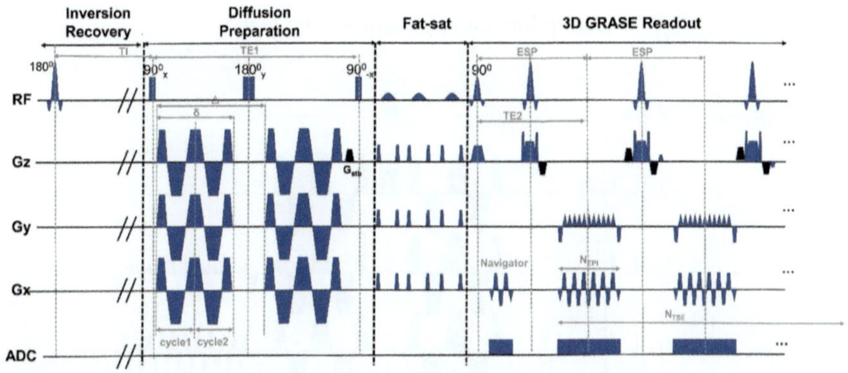

Fig. 6.1 Schematic diagram of the 3D IR-OGprep-GRASE sequence framework

a suitable TI before the sequence undergoes diffusion weighting. In this study, a 3D IR-PGprep-GRASE sequence was also designed for acquiring TDDMRI signals at long diffusion times, differing from the 3D IR-OGprep-GRASE sequence only in the different gradients added in the diffusion preparation module. Due to the limitation of the TR time in the IR technique, only single-segment acquisitions were used in this study.

The signal equation in the IR module [12] is shown in Eq. 6.1:

$$S = k\rho\left(1 - 2e^{-TI/T1} + e^{-TR/T1}\right)e^{-TE/T2} \tag{6.1}$$

This equation describes the decay and recovery process of the signal in a sequence acquisition, where S is the acquired signal, k is a scaling factor, and ρ is the spin density. The exponential term in the equation describes the influence of the T_1 and T_2 decay effects, with T_1 representing the longitudinal relaxation time of the tissue and T_2 representing the transverse relaxation time of the tissue. Selection of a suitable TI can realize the suppression of CSF signal. In this sequence, since the value of $T_{1\text{CSF}}$ is 4000 ms (at 3 T magnetic field strength) and TR is 9000 ms according to the experimental conditions, the CSF signal can be brought close to zero by setting the TI to 2500 ms. In this way, the signal intensity of the CSF in the image will be greatly reduced, resulting in better display of other tissue structures and measurement of ADC values. In this study, the IR module was also added to the clinically used 2D OG-EPI sequence, as shown in Fig. 6.2, to test the potential value of the IR module in TDDMRI.

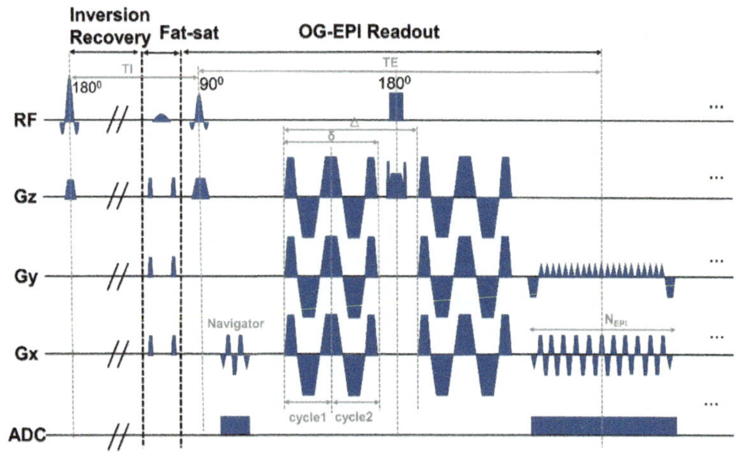

Fig. 6.2 Schematic diagram of the 2D IR-OG-EPI sequence framework

6.3 Experimental Methods

6.3.1 MRI Experiments

In this study, 3D GRASE and 2D EPI sequences were used to test the inhibitory effect of the IR module on CSF and its value in measuring TDDMRI. The 3D GRASE sequence was used for testing in healthy volunteers, while the 2D EPI sequence was applied to both healthy volunteers and brain tumor patients. Ten healthy volunteers (mean age = 22 ± 2 years, range 20–25 years, 6 males and 4 females) as well as six patients with brain tumors (mean age = 61 ± 8 years, range 48–71 years, 4 males and 2 females) were included in this study. All scanning experiments were performed on a 3 T Siemens MAGNETOM Prisma scanner equipped with a 64-channel head coil, which has a maximum gradient of 80 mT/m and a maximum gradient conversion rate of 200 mT/m/ms. A total of three sets of experiments were performed in this study. Experiments 1 and 2 were used to evaluate the effect of applying the IR module in 3D and 2D sequences, respectively, and to clarify whether the IR module has any application value under these two sequences. Experiment 3 further explored the inhibitory effect of the IR module on CSF in brain tumors, with the intention of revealing the potential value of the IR module for the clinical application of TDDMRI technology. Before conducting the experiments, all of them had fully understood the purpose and process of the study and signed the informed consent form. All study steps were approved by the ethical review board of the affiliated institutions.

6.3.2 Data Acquisition

Experiment 1: For healthy volunteers, experiments using diffusion-prepared 3D GRASE sequences were performed to assess the effect of using the IR module on diffusion time-dependent measurements. The dMRI signals were acquired using PG encoding ($\delta/\Delta = 10/40$ ms, 0 Hz, $\Delta_{eff} = 36.67$ ms) and OG encoding at 20 Hz (1 cycle, $\Delta_{eff} \approx 12.5$ ms), 40 Hz (2 cycles, $\Delta_{eff} \approx 6.25$ ms), 60 Hz (3 cycles, $\Delta_{eff} \approx 4.16$ ms) frequencies. At each diffusion time, other parameters were kept constant: FOV = 220 × 200 × 40 mm^3, resolution = 2 × 2 × 4 mm^3, TR/TI/TE_1/TE_2 = 9000/2500/103.46/31.94 ms, $N_{EPI} = 39$ (partial Fourier factor = 78%), $N_{FSE} = 12$, $b = 420$ s/mm^2. Six diffusion directions were acquired at each diffusion time and scans were repeated twice, and $b = 0$ was repeated four times for signal averaging. This experiment was acquired using a 3D GRASE sequence with single-segment excitation, with the GRAPPA acceleration factor along the EPI phase-encoding direction set to 2. The FOV was centered on the hippocampus and lateral ventricles. The scan time at each diffusion time with or without the IR module applied was 1.8 min.

Experiment 2: For healthy volunteers, experiments using diffusion-weighted 2D EPI sequences were performed to assess the effect of using the IR module on diffusion time-dependent measurements. The diffusion encoding scheme and imaging parameters were matched to the 3D sequence except for TR/TI/TE = 5000/2000/135 ms and single excitation $N_{EPI} = 100$. According to Eq. 4.1, TI was set to 2000 ms when TR was set to 5000 ms. TE was the same as $TE_1 + TE_2$ in Experiment 1. The scanning time at each diffusion time with or without the IR module applied was 1 min.

Before the start of Experiments 1 and 2, ADCs at different diffusion times in 3D GRASE and 2D EPI sequences with and without the IR module were first tested in an agar gel body model using the same sequence and acquisition parameters.

Experiment 3: For patients with gliomas, experiments were performed using diffusion-weighted 2D EPI sequences to assess the diffusion time-dependent effects of the applied and unapplied IR modules. PG-encoded dMRI was performed using $\delta/\Delta = 10/30$ ms, 0 Hz, $\Delta_{eff} = 26.67$ ms. og-encoded dMRI was acquired at 25 (1 cycle, $\Delta_{eff} \approx 10$ ms) and 50 Hz (2 cycles, $\Delta_{eff} \approx 5$ ms) were acquired. Other parameters of the scans were kept constant: FOV = 220 × 220 mm^2, resolution = 2.8 × 2.8 mm^2, 10 layers with a layer thickness of 4 mm, TR/TI/TE = 5000/2000/120 ms, $N_{EPI} = 80$, b-value of 500 s/mm^2, and the GRAPPA acceleration factor was set to 2. Scan time at each diffusion time with or without the IR module applied was 1 min.

6.3.3 Data Analysis

The k-space data obtained from the 3D GRASE sequence were first reconstructed by the GRAPPA technique with Tikhonov regularization. The reconstructed images were then subjected to several pre-processing procedures using MRtrix software in this experiment, including: Gibbs artifact removal, eddy current correction, image alignment and bias field correction. In the experiments with healthy volunteers (Experiments 1 and 2), multiple ROIs such as hippocampal subregions (head, body, and tail), thalamus, and hindbrain white matter were manually outlined as shown in Fig. 6.4a. In the experiment with glioma patients (Experiment 3), the tumor region and contralateral healthy tissues were outlined layer by layer, as shown in Fig. 6.7a. To ensure the professionalism and accuracy of the results, all ROIs were manually outlined by a neuroradiologist with 5 years of experience.

Experiments 1 and 2 were analyzed using MATLAB software with ANOVA and post hoc t-tests to assess differences in ADC at different diffusion times. Experiment 1 also used a linear regression model to assess the rate of change in ADC associated with diffusion time and an analysis of covariance (ANCOVA) was performed using SPSS (Statistics for Windows, Version 25.0, IBMCorp., Armonk, N.Y., USA) software to assess the variability in regression slopes. Experiment 3 used a linear regression model for 2D EPI sequences with and without the IR module applied to assess the rate of ADC change in relation to diffusion time in tumor and contralateral healthy tissue. For all results, the Bonferroni correction of $p < 0.05$ was considered statistically different. Meanwhile, the IMPULSED biophysical model was used in this experiment to estimate the tumor intracellular fraction (f_{in}), cell diameter (d), and extracellular diffusivity (D_{ex}) by fixing the intracellular diffusivity at 1×10^{-3} mm^2/s and assuming that the extracellular diffusion is not affected by diffusion time. Curves were fitted according to the least squares method using MATLAB software and repeated 100 times with random initialization to avoid local minima.

6.4 Experimental Results

Figure 6.3 demonstrates the ADC metrics in 3D GRASE and 2D EPI sequences in agar gel body molds with and without the IR module applied at different diffusion times, and the results show that there is no significant sequence difference in ADC at any diffusion time ($p > 0.1$). Since agar gel body models are uniformly textured objects, the experimental results were as expected, i.e., no difference in ADC at any diffusion time.

Figure 6.4 demonstrates the b_0 images and calculated ADC images obtained using the 3D GRASE sequence at different diffusion times with and without the IR module applied. In both b_0 images and DWI images (Fig. 6.4a–d), the IR module has a good suppression effect on the CSF signals in the ventricular and sulcal gyrus, which is important for mitigating the effect of CSF on ADC measurements. In both datasets it was shown that the ADC increased with increasing oscillation frequency (e.g., Fig. 6.4e and f).

Figure 6.5 further reveals a quantitative comparison of the ADC diffusion time dependence with and without the IR module applied. It is clear that the diffusion time-dependent changes of ADC in the thalamic and hindbrain white matter regions are significant both in the 3D GRASE sequence with and without the IR module

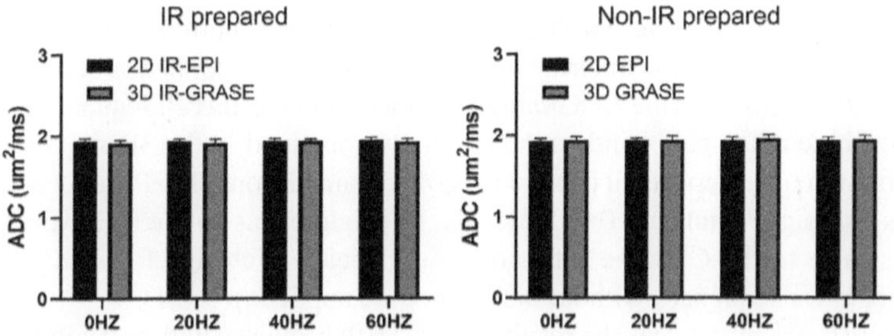

Fig. 6.3 Diffusion time dependence of ADC measured by 3D GRASE versus 2D EPI sequences in a body model. There was no significant difference in ADC values between 2 and 3D sequences with or without the IR module applied at 0 (PG), 20, 40 and 60 Hz

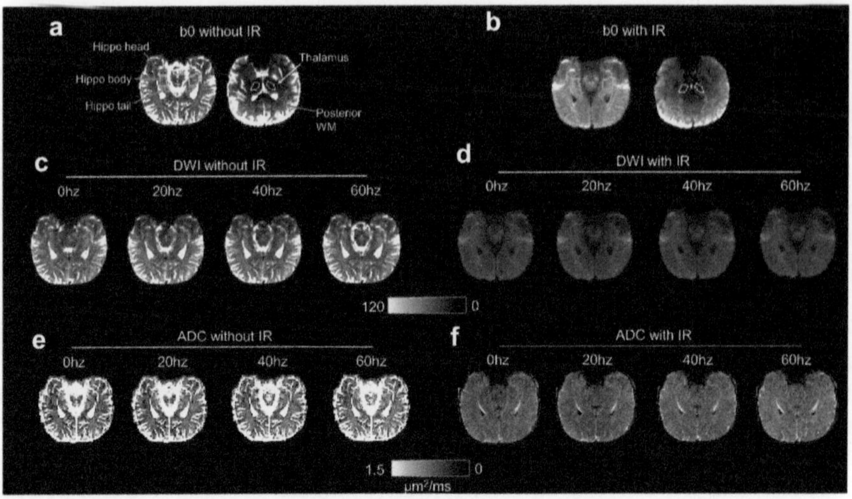

Fig. 6.4 b_0 and ADC images obtained for the 3D GRASE sequence with and without the IR module applied. **a** and **b** 3D GRASE sequences acquired in b_0 images acquired with or without the IR module applied. Regions of interest outlined in the b_0 images, including hippocampal head, body and tail, thalamus and hindbrain white matter. **c–f** DWI and ADC maps acquired at 0 (PG), 20, 40 and 60 Hz with and without IR module (**d**, **f**) applied to the 3D GRASE sequence (**c**, **e**). *Abbreviations* hippo—hippocampus; WM—white matter [33]

applied. Whereas in the hippocampal (including head, body, and tail) regions close to the CSF, the sequences without the IR module imposed were not significant in terms of diffusion time-dependent changes in ADC (p-values of 0.68, 0.37, and 0.87, respectively), the diffusion time-dependent changes in ADC in these brain regions became significant after the IR module was imposed ($p < 0.05$). Post hoc t-tests showed that the ADCs between individual diffusion times were significantly different in all brain regions scanned with the IR module applied ($p < 0.01$), and that the ADCs measured for the sequences with the IR module applied were significantly lower than those for the sequences without the IR module applied ($p < 0.01$).

Figure 6.6 illustrates the ADC values obtained from the 2D EPI sequence measurements. These data again validate the results we obtained in the 3D sequences, i.e., the diffusion time-dependent change in ADC obtained from 2D EPI sequence acquisitions was significant ($p < 0.01$) after the IR module was applied. However, in the region close to the CSF, the diffusion time-dependent change for ADCs was not significant for sequences without the IR module applied ($p > 0.1$).

Figure 6.7 demonstrates the results of a 2D EPI sequence with and without the IR module applied to scan six glioma patients and measure the diffusion time dependence of the ADC. Figure 6.7b shows that the IR module had a slight effect on the contrast of the tumor in the DWI images. Further quantitative analysis confirmed that the diffusion time dependence of ADC in the tumor region was significantly higher than that in the contralateral healthy tissue region both with and without the IR module applied to the sequence ($p < 0.01$). Furthermore, in the tumor region, the slope of the diffusion time dependence of ADC obtained for sequences with the IR module applied (slope = 0.0196 um^2/ms^2) was significantly higher than for sequences without the IR module applied (slope = 0.0034 um^2/ms^2) ($p < 0.01$). However, for the contralateral healthy tissue region, the diffusion time dependence of ADCs did not show a clear difference regardless of whether the IR module was applied or not ($p = 0.52$), as shown in Fig. 6.7d. Such a result may stem from the fact that the IR module had a significant inhibitory effect on water molecule signals in the tumor region, whereas it did not have the same degree of inhibitory effect on the contralateral tissues, especially those regions where edema or necrosis may be present. This finding highlights the potential value and important role of the IR module in enhancing tumor contrast and improving ADC diffusion time-dependent measurements.

In this experiment, microstructural parameters (tumor intracellular fraction, cell diameter, and extracellular diffusivity) were estimated based on the IMPULSED model for six glioma patients, and the cellular microstructural parameters (tumor intracellular fraction, cell diameter, and extracellular diffusivity) were obtained using 2D EPI sequences with and without the IR module applied, respectively (Fig. 6.8a and b). The regional distributions of tumor intracellular fraction and extracellular diffusivity acquired by the non-IR sequence and the IR sequence were similar, but the diameter distributions were different between the two. The difference plot of microstructural parameters (Fig. 6.8c) showed that the values of microstructural parameters fitted using IR sequences were all higher than non-IR sequences. This further validates that the IR module improves the diffusion time dependence of ADCs.

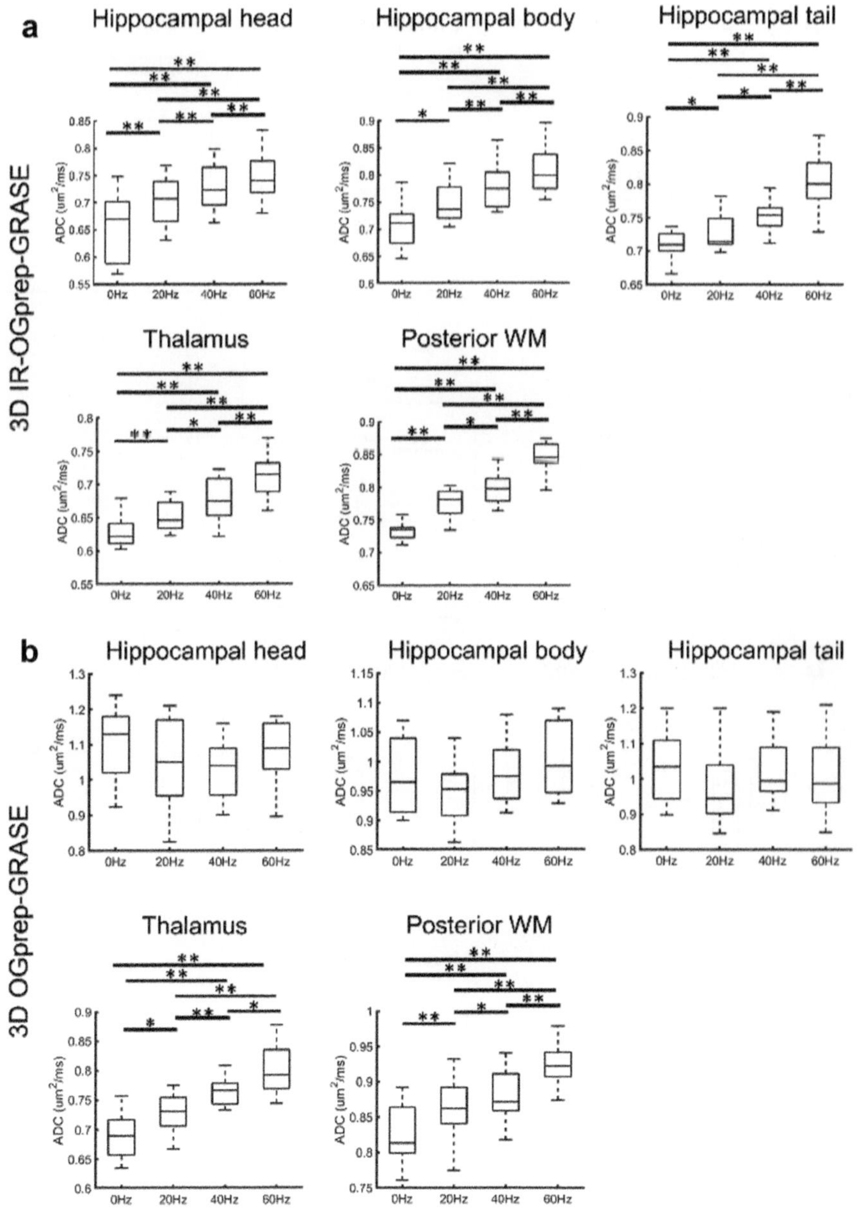

Fig. 6.5 Diffusion time dependence of ADC measured by 3D GRASE sequence at different diffusion times. Measurements were performed using the 3D IR-OGprep-GRASE and 3D IR-PGprep-GRASE (**a**), 3D OGprep-GRASE and 3D PGprep-GRASE (**b**) sequences. For sequences with the IR module applied, all ROIs showed significant diffusion time-dependent changes in ADC. In contrast, the hippocampal (head, body, and tail) regions close to the ventricles and sulci did not show significant diffusion time-dependent changes on ADC in the absence of the imposed IR module (*$p < 0.05$ and **$p < 0.01$ analyzed by ANOVA and post hoc t-tests) [33]

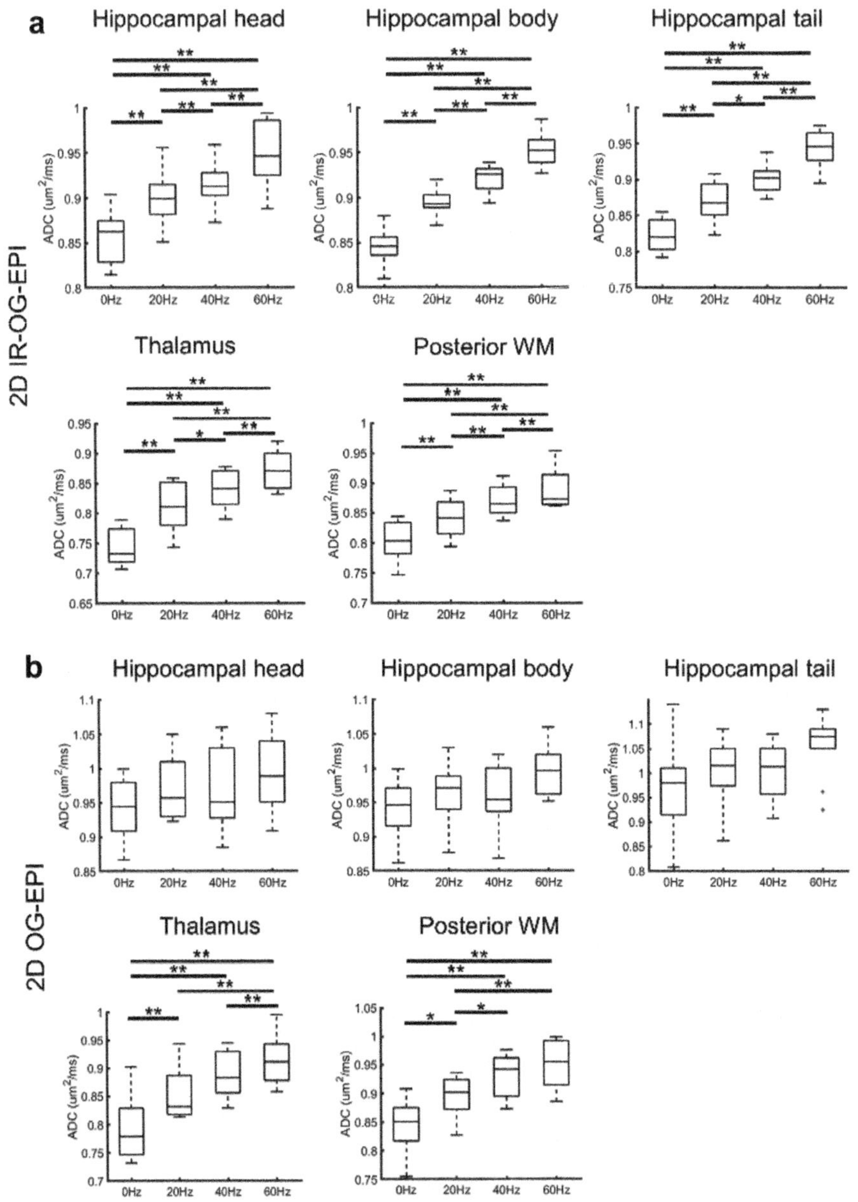

Fig. 6.6 Diffusion time dependence of ADCs measured by 2D EPI sequences at different diffusion times. Measurements were performed using 2D IR OG EPI and 2D IR PG-EPI (**a**), and 2D OG-EPI and PG-EPI (**b**) sequences. For sequences with the IR module applied, all ROIs showed significant diffusion time-dependent changes in ADC. In contrast, the hippocampal (head, body, and tail) regions close to the ventricles and sulci did not show significant diffusion time-dependent changes on the ADC without the IR module imposed (*$p < 0.05$ and **$p < 0.01$ by ANOVA and post hoc *t*-test)

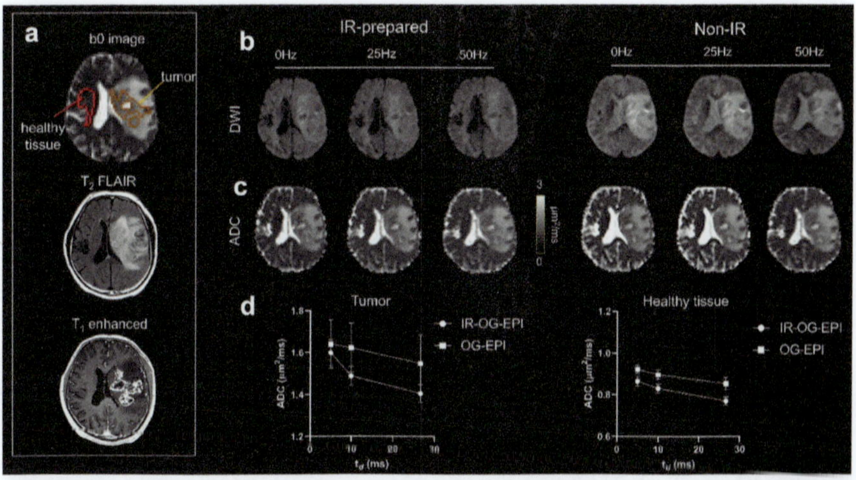

Fig. 6.7 Diffusion time dependence of ADCs in the brains of glioma patients measured by 2D EPI sequences. **a** T2-weighted, T2 FLAIR, and T1-enhanced images of a glioma patient. Areas of tumor and contralateral healthy tissue are depicted on the b_0 image with reference to the T2 FLAIR and T1 enhanced images. **b** DWI maps and **c** ADC maps obtained from 2D EPI sequences acquired with and without the IR module applied at multiple oscillation frequencies including 0, 25 and 50 Hz. **d** Measurement of the diffusion time dependence of ADC in all six subjects in both regions. The ADCs of both sequences showed significant diffusion time-dependent changes, but the sequence with the IR module applied had a higher slope from 5 (50 Hz) to 10 ms (25 Hz) in the tumor region ($p < 0.05$) [33]

6.5 Discussion

In this study, 3D IR-OGprep-GRASE and 3D IR-PGprep-GRASE sequences designed for TDDMRI are presented. Compared with the 3D OGprep-GRASE and 3D PGprep-GRASE sequences developed in the previous chapters, this sequence has a new IR module. It effectively inhibits CSF signaling and thus improves the time-dependence of diffusion parameters in brain tissues close to the ventricles and sulcal gyrus, e.g., the hippocampus. It is noteworthy that this property was observed in this study not only in the 3D GRASE sequence, but also the same result was found in the 2D EPI sequence. This suggests that the introduction of an IR module is necessary when performing TDDMRI measurements, regardless of whether 2D or 3D acquisition is used. In addition, preliminary results obtained in patients with gliomas further validate the potential value of using IR technology for TDDMRI measurements.

IR techniques have been proposed in numerous dMRI studies, mainly for suppression of cerebrospinal fluid signals, thus improving the accuracy of diffusion parameters [13, 14] as well as the performance of fiber bundle tracking [15, 16]. For example, a study by Chen et al. [16] confirmed that fiber bundle volumes measured

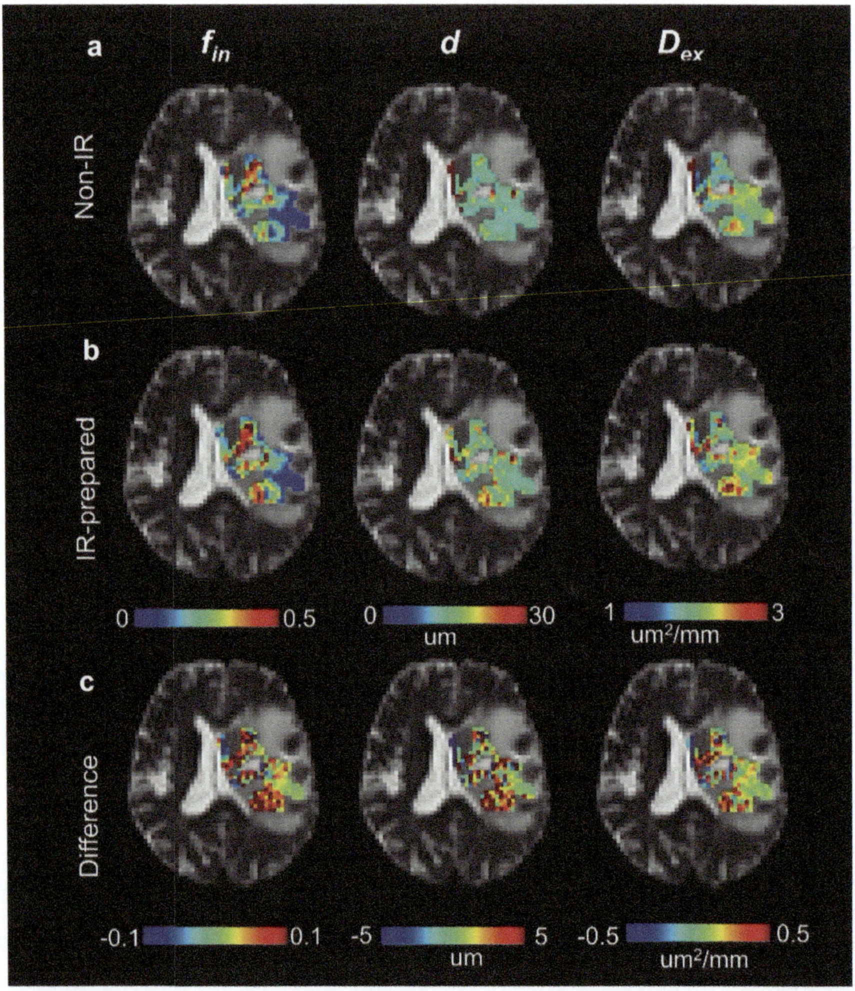

Fig. 6.8 2D EPI sequences fitting microstructural parameters of the brain of glioma patients based on the IMPULSED model. Fitting of 2D EPI sequences without applying (**a**) or with applying (**b**) the IR module yields the corresponding mapping of microstructural parameters such as f_{in}, d, and D_{ex} to the b_0 image. **c** Difference maps (IR-non-IR) of the microstructural parameter maps

using IR-DTI were 9.7% larger than those obtained by conventional DTI measurements. Xu et al. [14] used an IR module in 2D EPI to reduce artifacts produced by CSF in axon diameter imaging of the corpus callosum. However, that study did not assess the specific effects of CSF inhibition. In the present study, more severe signal attenuation may have resulted due to the use of the long readout module of 3D GRASE [11]. Compared to the 2D OG-EPI sequence, the CSF scanned with the 3D OGprep-GRASE sequence may exhibit higher PVE.

Although the IR module proposed in this study showed effectiveness in suppressing the signal of CSF, the TR had to be prolonged in order for the tissue signal to be adequately recovered, which undoubtedly increased the length of the scan. Therefore, in addition to the use of IR techniques, there is a need to explore other possible alternative countermeasures against the CSF effect, both at the level of data acquisition and post-processing. For example, many studies [17–19] have used *b*-values greater than zero to exclude CSF signals, thus allowing for more accurate calculation of diffusion parameters. Specifically, Baron and Beaulieu [18] used a minimum value of *b* that is not zero ($b_{\min} = 425$ s/mm^2) and a relatively short TR (3 s) to improve the accuracy of the DTI fiber tracking. However, the application of this method can be limited due to the limited *b*-value realized by the clinical scanner in OG-dMRI. Recently, Wu and Zhang [20] proposed a dual-frequency diffusion encoding technique using an additional PG gradient orthogonal to the OG gradient to suppress microcirculatory flow. In addition, tissue water and CSF signals can also be separated by fitting using compartmentalized chamber models, e.g., fitting the tissue tensor and isotropic CSF tensor to dMRI data [21, 22]. However, in order to be able to do an accurate fit, dMRI data need to be acquired for multiple diffusion directions and *b*-values, which will undoubtedly also increase the acquisition time of TDDMRI.

The 3D IR-OGprep-GRASE and 3D IR-PGprep-GRASE sequences used in this study measured the diffusion time dependence of ADC in brain regions. Among them, this study focused on the diffusion time dependence in different subregions of the human brain in the hippocampus. As a key structure in the brain, the hippocampus has a crucial role in learning and memory processes. Notably, the hippocampus is not a simple whole but a complex structure composed of multiple subregions. Early studies have shown significant morphological and functional differences between hippocampal subregions [23, 24]. For example, a study by Ly et al. [25] found significant differences in microstructural parameters, such as MD, RD, AD, and FA, among different subregions of the hippocampus. Therefore, the present study hypothesized that the differences in microstructural organization of hippocampal subregions may lead to differences in their diffusion time-dependence. In the present study, we found that the ADC values in the head of the hippocampus were relatively low, and the ADC increased at a faster rate with increasing diffusion time. In contrast, in the hippocampal body and tail, the ADC growth rate was slower. This result suggests that there is a significant difference between the hippocampal head and the rest of the subregions in the diffusion time dependence of ADC, which may stem from the high cell density [26, 27] of the hippocampal head and lead to the difference in microstructure.

In addition to the study of diffusion time-dependence of ADCs in the hippocampal subregion, the present study has given equal attention to the diffusion time-dependence of ADCs in gliomas. Earlier studies have revealed a close correlation between changes in brain tumor microstructure and ADC diffusion time dependence [28]. On the basis of this correlation, various models, such as IMPULSED and POMACE, have been proposed to describe tumor microstructure [6, 29]. The feasibility of these models has also been demonstrated in brain studies [30, 31]. In this study, we further investigated the effect of CSF inhibition on TDDMRI of brain tumors and found that the diffusion time-dependence of ADC of brain tumors could be observed more precisely by inhibiting CSF signals. In this experiment, the IR technique was used to achieve CSF signal suppression, and the results obtained showed that the diffusion time dependence of ADC in the brain tumor region was significantly increased by the addition of the IR module. This phenomenon suggests that the intracellular fraction or cell density may increase after effective suppression of CSF signaling in edematous or necrotic regions surrounding the tumor [7, 28]. This may help in brain tumor diagnosis, treatment and prognostic analysis, etc. However, in order to further test this hypothesis, more extensive clinical studies are still needed to recruit more patients for validation. In addition, this study also compared the effects of sequences with and without the IR module applied on the diffusion time-dependence of ADCs in healthy tissues contralateral to brain tumors. It was found that there was no significant difference in the diffusion time dependence of ADC in contralateral healthy tissues between the two sequences. This further suggests that the increase in ADC diffusion time dependence induced by the IR module may originate from the inhibition of water signaling in tumor-specific edema and necrotic regions. According to previous studies, the T_1 value of edematous tissues is not a constant value, but varies in the range of 1000–3000 ms [32]. Therefore, it can be speculated that water molecule signaling in edematous regions may be partially inhibited by the IR module.

Because of the relatively long scan time required for TDDMRI imaging involving the IR module, there are two limitations to this study. First, only a limited number of tumor patients were recruited for this preliminary study to demonstrate the inhibitory effect of the IR module on CSF. To more directly demonstrate the clinical value of IR technology in tumor detection, the number of patients needs to be further increased. Then, in order to avoid excessive scanning time, this study used a single-segment acquired 3D GRASE sequence, which can only acquire data on relatively thin layers (10 layers). Nevertheless, the preliminary results of this study still provide a new perspective and research methodology for studying and understanding the diffusion time dependence of brain tumor ADC. These findings are expected to optimize neuroimaging measurements and provide a more accurate basis for clinical diagnosis, as well as provide important references and guidance for a deeper understanding of brain structure and function.

6.6 Conclusion

The 3D IR-OGprep-GRASE and IR-PGprep-GRASE sequences proposed in this chapter effectively suppressed the effect of CSF and significantly improved the accuracy of measuring ADC diffusion time-dependence in the brain, especially in regions adjacent to CSF. In addition, the IR module also embodies a crucial role for the accurate measurement of the diffusion time dependence of ADC in tumor regions, providing a theoretical basis for the potential application of TDDMRI in clinical practice. Overall, the sequences presented in this chapter provide new possibilities for researchers to further understand and study in depth the microstructure and function of tissues close to the CSF region.

References

1. Xu J, Li H, Harkins KD et al (2014) Mapping mean axon diameter and axonal volume fraction by MRI using temporal diffusion spectroscopy. Neuroimage 103:10–19
2. Drobnjak I, Zhang H, Ianuş A et al (2016) PGSE, OGSE, and sensitivity to axon diameter in diffusion MRI: insight from a simulation study. Magnet Reson Med 75(2):688–700
3. Tétreault P, Harkins KD, Baron CA et al (2020) Diffusion time dependency along the human corpus callosum and exploration of age and sex differences as assessed by oscillating gradient spin-echo diffusion tensor imaging. Neuroimage 210:116533
4. Baron CA, Beaulieu C (2014) Oscillating gradient spin-echo (OGSE) diffusion tensor imaging of the human brain. Magnet Reson Med 72(3):726–736
5. Wu D, Liu D, Hsu YC et al (2020) Diffusion-prepared 3D gradient spin-echo sequence for improved oscillating gradient diffusion. MRI 85(1):78–88
6. Jiang X, Li H, Xie J et al (2017) In vivo imaging of cancer cell size and cellularity using temporal diffusion spectroscopy. Magnet Reson Med 78(1):156–164
7. Reynaud O, Winters KV, Hoang DM et al (2016) Pulsed and oscillating gradient MRI for assessment of cell size and extracellular space (POMACE) in mouse gliomas. NMR Biomed 29(10):1350–1363
8. Latour LL, Warach S (2002) Cerebral spinal fluid contamination of the measurement of the apparent diffusion coefficient of water in acute stroke. Magnet Reson Med Off J Int Soc Magnet Reson Med 48(3):478–486
9. Ni J-M, Mogensen MA, Chen Z-A et al (2010) Influence of fluid-attenuated inversion-recovery on stroke apparent diffusion coefficient measurements and its clinical application. Eur J Radiol 75(2):e76–e81
10. Hajnal JV, Bryant DJ, Kasuboski L et al (1992) Use of fluid attenuated inversion recovery (FLAIR) pulse sequences in MRI of the brain. J Comput Assist Tomogr 16:841–841
11. Wu D, Liu D, Hsu YC et al (2020) Diffusion-prepared 3D gradient spin-echo sequence for improved oscillating gradient diffusion MRI. Magnet Reson Med 85(1):78–88
12. Bydder GM, Young IR (1985) MR imaging: clinical use of the inversion recovery sequence. J Comput Assist Tomogr 9(4):659–675
13. Ma X, Kadah YM, Laconte SM et al (2004) Enhancing measured diffusion anisotropy in gray matter by eliminating CSF contamination with FLAIR. Magnet Reson Med Off J Int Soc Magnet Reson Med 51(2):423–427
14. Xu J, Li H, Li K et al (2016) Fast and simplified mapping of mean axon diameter using temporal diffusion spectroscopy. NMR Biomed 29(4):400–410
15. Chou M-C, Lin Y-R, Huang T-Y et al (2005) FLAIR diffusion-tensor MR tractography: comparison of fiber tracking with conventional imaging. Am J Neuroradiol 26(3):591–597
16. Cheng Y-W, Chung H-W, Chen C-Y et al (2011) Diffusion tensor imaging with cerebrospinal fluid suppression and signal-to-noise preservation using acquisition combining fluid-attenuated inversion recovery and conventional imaging: comparison of fiber tracking. Eur J Radiol 79(1):113–117
17. Salminen LE, Conturo TE, Laidlaw DH et al (2016) Regional age differences in gray matter diffusivity among healthy older adults. Brain Imag Behav 10(1):203–211
18. Baron CA, Beaulieu C (2015) Acquisition strategy to reduce cerebrospinal fluid partial volume effects for improved DTI tractography. Magnet Reson Med 73(3):1075–1084

19. Salminen LE, Conturo TE, Bolzenius JD et al (2016) Reducing CSF partial volume effects to enhance diffusion tensor imaging metrics of brain microstructure. Technol Innov 18(1):5
20. Wu D, Zhang J (2017) The effect of microcirculatory flow on oscillating gradient diffusion MRI and diffusion encoding with dual-frequency orthogonal gradients (DEFOG). Magnet Reson Med 77(4):1583–1592
21. Metzler-Baddeley C, O'sullivan MJ, Bells S et al (2012) How and how not to correct for CSF-contamination in diffusion MRI. Neuroimage 59(2):1394–1403
22. Pasternak O, Shenton ME, Westin C-F (2012) Estimation of extracellular volume from regularized multi-shell diffusion MRI. In: International conference on medical image computing and computer-assisted intervention, pp 305–312
23. Robinson JL, Barron DS, Kirby LA et al (2015) Neurofunctional topography of the human hippocampus. Hum Brain Map 36(12):5018–5037
24. De Flores R, Berron D, Ding SL et al (2020) Characterization of hippocampal subfields using ex vivo MRI and histology data: lessons for in vivo segmentation. Hippocampus 30(6):545–564
25. Ly M, Foley L, Manivannan A et al (2020) Mesoscale diffusion magnetic resonance imaging of the ex vivo human hippocampus. Hum Brain Map 41(15):4200–4218
26. Bernasconi N, Bernasconi A, Caramanos Z et al (2003) Mesial temporal damage in temporal lobe epilepsy: a volumetric MRI study of the hippocampus, amygdala and parahippocampal region. Brain 126(2):462–469
27. Dam AM (1980) Epilepsy and neuron loss in the hippocampus. Epilepsia 21(6):617–629
28. Jiang X, Li H, Xie J et al (2016) Quantification of cell size using temporal diffusion spectroscopy. Magnet Reson Med 75(3):1076–1085
29. Reynaud O, Winters KV, Hoang DM et al (2016) Pulsed and oscillating gradient MRI for assessment of cell size and extracellular space (POMACE) in mouse gliomas. NMR Biomed 13:11454
30. Xu J, Jiang X, Li H et al (2020) Magnetic resonance imaging of mean cell size in human breast tumors. Magn Reson Med 83(6):2002–2014
31. Wu D, Northington FJ, Fieremans E et al (2017) Diffusion time dependence of kurtosis reveals microstructural changes after neonatal hypoxia-ischemia
32. Andersen C, Haselgrove J, Doenstrup S, et al. Resorption of peritumoural oedema in cerebral gliomas during dexamethasone treatment evaluated by NMR relaxation time imaging. Acta Neuroch 993, 122(3): 218–224
33. Li H, Zu T, Hsu Y C et al (2023) Inversion-recovery-prepared oscillating gradient sequence improves diffusion-time dependency measurements in the human brain [J]. J Magn Reson Imaging 57(2): 446–453

Chapter 7
Summary and Outlook

Abstract Time-dependent diffusion MRI (TDDMRI) is a powerful tool for probing tissue microstructure, but it faces challenges such as low signal-to-noise ratio (SNR) and resolution, motion-induced artifacts, and cerebrospinal fluid (CSF) contamination. This study aims to address these critical issues by developing a suite of novel 3D whole-brain TDDMRI techniques. First, to overcome low SNR and resolution, a 3D oscillating gradient preparation (OGprep) GRASE sequence was developed, which significantly improved image quality and acquisition efficiency compared to conventional 2D EPI methods. Second, to enable high-fidelity, whole-brain multi-segment acquisition, dual-navigator and subsequently 3D-navigator modules were integrated to correct for inter-segment phase errors caused by motion, complemented by GRAPPA parallel imaging to reduce scan time. Finally, to eliminate CSF signal contamination, an inversion recovery (IR) module was incorporated (3D IR-OGprep-GRASE), enhancing the accuracy of diffusion measurements in adjacent tissues like the hippocampus and in tumors. Validated on a clinical 3 T MRI scanner, these combined techniques successfully achieved high-resolution (1.5 mm isotropic) whole-brain TDDMRI, effectively suppressed artifacts, and provided more accurate microstructural parameter measurements than standard methods. This work establishes a robust and clinically viable imaging protocol that overcomes major limitations of TDDMRI, paving the way for its broader application in neuroscience and clinical diagnostics.

Keywords Time-dependent diffusion MRI (TDDMRI) · 3D GRASE · Oscillating gradient · Motion correction · Navigator · Cerebrospinal fluid (CSF) suppression · Inversion recovery (IR)

7.1 Summary

As a widely used clinical imaging technique, dMRI continues to advance the field of medicine. Through a non-invasive imaging modality, dMRI is able to reveal microstructural features of biological tissues, which is important for the diagnosis

D. Wu et al., *Development and Application of 3D Diffusion Magnetic Resonance Imaging Sequences*, Advanced Topics in Science and Technology in China 5,
https://doi.org/10.1007/978-981-95-0808-2_7

and treatment of diseases such as stroke, tumors, epilepsy, and so on. However, until today dMRI still faces certain challenges. First, the excessive length of the TE and TR of the scan leads to low resolution and SNR of the image, which negatively affects the image quality and the accuracy of experimental results. Second, during multi-segment acquisition, phase errors generated by motion between different segments can cause severe image aliasing artifacts. Finally, the effect of CSF on the signal value of the surrounding tissue can lead to errors in the measurement results. These problems are particularly serious in TDDMRI. Therefore, how to improve imaging techniques and thus solve these problems is an academic challenge for both current and future research.

First, this study proposes a 3D OGprep-GRASE sequence to address the problem of low resolution and SNR of images due to the excessively long TE and TR of TDDMRI scans at short diffusion times. In this study, the imposition of the diffusion gradient of the conventional TDDMRI sequence and the acquisition of the k-space signals were effectively separated by an innovative sequence design. Then, to further enhance the acquisition efficiency, the acquisition module was designed as 3D GRASE encoded. Compared with the conventional 2D EPI encoding for TDDMRI sequences, the newly designed sequence enhanced the SNR and resolution of the images and significantly reduced the scan time. This study was also experimentally validated on a clinical 3 T MRI scanner. Thanks to the significant improvement in image SNR, this study achieved reliable diffusion tensor reconstruction of OG-dMRI on a clinical system and accurately characterized the diffusion time-dependent ADC changes in brain regions. The application of 3D OG-dMRI sequences opens up a new pathway for the conversion of TDDMRI under short diffusion time from preclinical to clinical applications and creates a greater value for future clinical applications.

Then, in order to solve the problem that the 3D GRASE readout module can only acquire in a single segment, this study constructed a 3D OGprep-GRASE and PGprep-GRASE sequence based on dual-navigator multi-segment acquisitions on the basis of the 3D OGprep-GRASE and PGprep-GRASE sequences. The dual-navigator module in the new sequences provided a phase reference for the signals acquired in each segment, thus correcting the phase error between multi-segment acquisitions and realizing TDDMRI whole-brain imaging. Meanwhile, the GRAPPA parallel acceleration technique used in this study in the EPI phase encoding direction of the 3D GRASE sequence not only reduces the TE, but also improves the contour of the PSF, further enhancing the image quality. In this study, a detailed comparison of dMRI data acquired at the same scan time between a dual-navigator-based multi-segment acquisition of 3D GRASE sequences and conventional 2D EPI sequences on a clinical 3 T MRI scanner was performed, and the results showed that the 3D sequences had a significant advantage in terms of SNR. In addition to this, the present study further compared the fiber tracking results and microstructural parameters of 3D and 2D sequences at the same SNR, and found that in the corticospinal tracts, 3D sequences had higher fiber bundle cross-sectional area and richer fiber details under the cerebral cortex. Therefore, 3D OGprep-GRASE and PGprep-GRASE sequences based on multi-segment acquisitions with dual navigators play a very critical role in discovering and accurately reflecting brain fibers and microstructural parameters, and provide an important tool for future clinical studies and applications. Next, this study proposed 3D OGprep-GRASE and PGprep-GRASE sequences for multi-segment acquisitions based on 3D navigator, based on the current situation that the 2D navigator has limited effect of phase correction in z-direction. The 3D navigator module further corrects the inter-segment phase error and improves the imaging quality. Meanwhile, as more diffusion time points in the acquisition would lead to long scanning time, 3D GRAPPA acceleration was further introduced in this study. Compared with 2D GRAPPA, 3D GRAPPA not only reduces TE, but also reduces TR. With this sequence, this study realized 1.5 mm isotropic high-resolution TDDMRI whole-brain imaging on a clinical 3 T magnetic resonance scanner, which revealed different diffusion time-dependence based on a power-law model in different regions of the cerebral cortex. In this study, theta was found to be higher in the cerebral sensory cortex and lower in the higher cortex. Subsequently, by analyzing the correlation of θ with microstructural parameters obtained from NODDI and DKI model fitting, the present study found that complex microstructure corresponded to higher neuronal synapse density, higher directional dispersion, and non-Gaussian diffusion. Thus, 3D OGprep-GRASE and PGprep-GRASE sequences based on multi-segment acquisitions with 3D navigator not only provide a new scheme for improving SNR and imaging quality of TDDMRI images, but also provide important support for understanding diffusion time-dependent microstructural parameters based on diffusion as well as for related analytical and applied studies.

Finally, to address the problem of CSF affecting the signal values of surrounding tissues in TDDMRI images, this study further developed 3D IR-OGprep-GRASE and IR-PGprep-GRASE sequences based on 3D OGprep-GRASE and PGprep-GRASE sequences, in which the introduced IR module suppressed the CSF effect that enabling more accurate measurement of brain microstructure information. When scanning healthy subjects and tumor patients on a clinical 3 T MRI scanner using the sequences with the added IR module, this study found that the module has a significant advantage in measuring the diffusion time dependence of the brain, especially in the hippocampal region surrounded by CSF. In addition to this, the IR module is also important for accurately measuring diffusion time dependence in tumor regions. Therefore, the 3D IR-OGprep-GRASE and IR-PGprep-GRASE sequences can more accurately obtain the diffusion parameters of brain microstructures around CSF, which has significant impact and value in TDDMRI research and has a broad application prospect for the diagnosis and treatment of diseases.

Combining the above work, this study solved a series of key problems such as long scanning time, low image resolution and SNR, phase error caused by motion in multi-segment acquisition, and CSF affecting peripheral tissues, and realized a set of 3D whole brain novel TDDMRI techniques with high SNR, high resolution, and high-fidelity multi-segment acquisition. It is of great value in clinical applications. We recommend the imaging sequence and method combining these four research results in clinical TDDMRI imaging. This new scanning program can meet the high clinical standards. In our future work, we will continue to optimize this imaging technique so that it can play a more important role in basic neuroscience research and clinical applications.

7.2 Research Innovation

The main research innovations of this paper are summarized below.

Overall, this thesis provides sequences and technical solutions for high-quality acquisition of TDDMRI, and these results are expected to be able to advance TDDMRI acquisition techniques, as well as advance the use of TDDMRI imaging in clinical settings.

1. Aiming at the problems of low SNR and resolution, poor imaging quality and long acquisition time in TDDMRI imaging under short diffusion time with conventional 2D EPI encoding in clinical practice, the present study proposes a solution of replacing 2D EPI encoding with 3D GRASE encoding, i.e., 3D OGprep-GRASE sequence. This sequence not only realizes TDDMRI imaging under short diffusion time by OG encoding, but also reduces the acquisition time by gradient preparation technique. In addition, the 3D GRASE encoding shortens the scanning time and improves the image SNR and resolution. Meanwhile, the 3D GRAPPA parallel acceleration technique implemented on the 3D GRASE sequence further shortens the TE and image acquisition time, enhances the SNR, and realizes the goal of high-resolution TDDMRI imaging.
2. To address the problem that TDDMRI is sensitive to motion in multi-segment acquisition imaging, resulting in phase errors and aliasing artifacts. In this study, 3D OGprep-GRASE and PGprep-GRASE sequences of multi-segment acquisitions based on the navigator module were introduced for imaging. These sequences not only corrected the phase error due to motion between multi-segment acquisitions and eliminated the aliasing artifacts in the images, but also shortened the scanning time in general by reducing the need for rescanning and improved the efficiency of the clinical process. In this study, three methods, 1D, 2D, and 3D navigators, were progressively developed, and the progressively improved performance of these three methods in terms of phase error correction and removal of image aliasing artifacts was experimentally verified on a Siemens Prisma 3 T MRI machine.
3. To address the problem of CSF affecting the signal values of surrounding tissues in current TDDMRI imaging, this study introduced 3D IR-OGprep-GRASE and IR-PGprep-GRASE sequences based on the IR module for imaging. the IR module effectively suppressed the PVE and PSF of the CSF, and significantly improved ADC diffusion time in the region close to the CSF region and in the tumor region. dependent measurement accuracy. Therefore, the addition of the IR module is crucial for TDDMRI studies and related microstructural parameter estimation.

7.3 Outlook for Work

Based on the work of this study, there are the following directions that could be further pursued:

1. Research and development of 3D GRASE sequences based on multi-segment simultaneous acquisition. Although the 3D GRASE sequence for whole-brain multi-segment acquisition proposed in this study uses 3D GRAPPA parallel acceleration, the multi-segment acquisition technique still results in an imaging speed that cannot be further improved. Therefore, it is particularly important to develop a 3D GRASE sequence with multiple simultaneous acquisitions to further shorten the imaging time. However, it is worth noting that due to the complexity of imaging 3D GRASE sequences, the combination of these techniques and the corresponding image reconstruction methods need to be considered comprehensively.
2. Combining a 3D navigator-based imaging technique with a deep neural network-based motion correction method. The 3D navigator-based imaging technique proposed in this study can effectively correct the phase error of 3D GRASE between multiple acquisitions, but it may not be able to fully correct for motion artifacts if there is a wide range of head motion during acquisition. Deep neural network-based models can train images with motion artifacts to artifact-free images by pre-training the data, but they are not effective in solving the phase error between multi-segment acquisitions. Therefore, if the 3D navigator-based imaging technique and the deep neural network model-based approach are combined, the image artifacts in dMRI can be corrected more accurately.
3. Design and development of sequences for ultra-high-performance gradient systems and high-field TDDMRI imaging. In this study, the proposed 3D GRASE sequence for multi-segment acquisition based on 3D navigator realizes TDDMRI whole brain imaging at high resolution. The ultra-high-performance gradient system and ultra-high-field imaging with enhanced gradient and magnetic field performance not only shorten the time of gradient encoding, but also shorten the acquisition time of GRASE. The shorter TE not only gives the image higher resolution and SNR, but also reduces image artifacts. This means that the ultra-high-performance gradient system and ultra-high field can facilitate the acquisition of sub-millimeter resolution images for 3D GRASE sequences based on multi-segment acquisitions of 3D navigators, facilitating neuroscientific research related to brain regions that require ultra-high-resolution imaging, e.g., the cerebral cortex, the hippocampus, and other brain regions.
4. Realization of clinical application value. This study solves the problems currently encountered in TDDMRI and optimizes the image quality by implementing various sequence imaging techniques. Currently, this study experimentally analyzes brain images of healthy volunteers to preliminarily validate the effectiveness of these techniques and their advantages over conventional sequences for TDDMRI imaging. The fundamental purpose of this study is to enable OG imaging techniques to meet demanding clinical requirements and to promote the development of short diffusion time TDDMRI imaging techniques, but the application of sequences to clinically relevant diseases is limited in this paper. Therefore, in the future, the 3D OGprep-GRASE and PGprep-GRASE sequences

based on multi-segment acquisitions with 3D navigators can be extended to clinical MRI machines in more hospitals, so that the validity and stability of the sequences used can be tested in a wider range of clinical disease diagnosis.

GPSR Compliance

The European Union's (EU) General Product Safety Regulation (GPSR) is a set of rules that requires consumer products to be safe and our obligations to ensure this.

If you have any concerns about our products, you can contact us on ProductSafety@springernature.com

In case Publisher is established outside the EU, the EU authorized representative is:

Springer Nature Customer Service Center GmbH
Europaplatz 3
69115 Heidelberg, Germany

Batch number: 10755472

Printed by Printforce, the Netherlands